Lecture Notes in Computer Science　　7223

Commenced Publication in 1973
Founding and Former Series Editors:
Gerhard Goos, Juris Hartmanis, and Jan van Leeuwen

T0213162

Michael A. Lones   Stephen L. Smith
Sarah Teichmann   Felix Naef
James A. Walker   Martin A. Trefzer (Eds.)

Information Processing in Cells and Tissues

9th International Conference, IPCAT 2012
Cambridge, UK, March 31 – April 2, 2012
Proceedings

 Springer

Volume Editors

Michael A. Lones
Stephen L. Smith
James Alfred Walker
Martin Albrecht Trefzer
University of York, Department of Electronics
York, YO10 5DD, UK
E-mail: {mal503, sls5, jaw500, mt540}@ohm.york.ac.uk

Sarah Teichmann
MRC Laboratory of Molecular Biology
Hills Road, Cambridge, CB2 0QH, UK
E-mail: sat@mrc-lmb.cam.ac.uk

Felix Naef
École Polytechnique Fédérale de Lausanne (EPFL)
The Institute of Bioengineering
1015 Lausanne, Switzerland
E-mail: felix.naef@epfl.ch

ISSN 0302-9743   e-ISSN 1611-3349
ISBN 978-3-642-28791-6   e-ISBN 978-3-642-28792-3
DOI 10.1007/978-3-642-28792-3
Springer Heidelberg Dordrecht London New York

Library of Congress Control Number: 2012933437

CR Subject Classification (1998): J.3, F.1, F.2, I.5

LNCS Sublibrary: SL 1 – Theoretical Computer Science and General Issues

Typesetting: Camera-ready by author, data conversion by Scientific Publishing Services, Chennai, India

Printed on acid-free paper

Springer is part of Springer Science+Business Media (www.springer.com)

Preface

The 9th International Conference on Information Processing in Cells and Tissues took place from March 31 to April 2, 2012 at Trinity College, Cambridge. It followed previous events held in the Centro Stefano Franscini, Switzerland (2009), Oxford (2007), York (2005), Lausanne (2003), Brussels (2001), Indianapolis (1999), Sheffield (1997), and the inaugural IPCAT, which took place in Liverpool in 1995.

The original aim of IPCAT was to bring together a multidisciplinary group of scientists working on modelling cells and tissues, with a central theme of the nature of biological information and the ways it is processed in cells and tissues. Over the years, the conference has continued to attract scientists from many different disciplines, including biology, chemistry, computer science, electronics, mathematics, medicine and physics.

Computational modelling, in particular, has remained a prominent theme. This includes those using computational models and methods to understand biological systems, those developing new computational models and methods based on understanding of biological systems, and those applying these computational models and methods to problems in engineering, medicine and other fields.

For IPCAT 2012, we attempted to address the diversity of the IPCAT audience by assembling an Organizing Committee comprising biologists, computer scientists and engineers. To reflect the differing publication norms in different fields, this year we also gave authors the option of submitting either an extended abstract or a full paper, treating these equally during the review and ranking process. As a result of this, we accepted 13 full papers and 26 extended abstracts for presentation at the conference. Of these, 31 authors elected to have their work published in this volume of *Lecture Notes in Computer Science*.

To complement the technical program, we invited six renowned scientists to give keynote presentations. These each addressed particular aspects of information processing in biological cells and tissues:

- Madan Babu (MRC Laboratory of Molecular Biology, Cambridge, UK)
  "Intrinsically Disordered Segments and the Evolution of Protein Half-Life"
- Leonid A. Mirny (Massachusetts Institute of Technology, Cambridge, USA)
  "Higher-Order Chromatin Architecture: Bridging Physics and Biology"
- Karla Neugebauer (Max Planck Institute – CBG, Dresden, Germany)
  "Dynamics of Transcription and RNA Processing in Living Cells"
- Yitzhak Pilpel (Weizmann Institute, Israel)
  "Gene Expression Evolution Revealed from Lab Evolution Experiments"
- Pieter Rein ten Wolde (FOM Institute AMOLF, Amsterdam,
  The Netherlands)
  "Multiplexing Biochemical Signals"
- Anne-Claude Gavin (EMBL, Heidelberg, Germany)
  "Biological Networks from Proteins to Metabolites"

We would like to thank all the people involved in the organization and re-alization of IPCAT 2012, especially the authors, the invited speakers, and the members of the Program Committee, whose time and effort were central to the conference's success.

March 2012

Michael A. Lones
Stephen L. Smith
Sarah Teichmann
Felix Naef
James Alfred Walker
Martin Albrecht Trefzer

Organization

Organizing Committee

General Chairs

Stephen L. Smith	University of York, UK
Sarah Teichmann	Trinity College, Cambridge, UK
Felix Naef	École Polytechnique Fédérale de Lausanne, Switzerland

Program Chair

Michael A. Lones	University of York, UK

Publicity Chairs

Martin Albrecht Trefzer	University of York, UK
James Alfred Walker	University of York, UK

Program Committee

Alexander Bockmayr	Freie Universität Berlin, Germany
Rachel Cavill	Maastricht University, The Netherlands
Jerry Chandler	George Mason University, USA
Cristina Costa Santini	King Saud University, Saudi Arabia
Ron Cottam	Vrije Universiteit Brussel, Belgium
Antoine Danchin	CEA, France
Peter Erdi	Kalamazoo College, USA
Jianfeng Feng	University of Warwick, UK
Gary Fogel	Natural Selection Inc., USA
Jean-Louis Giavitto	IRCAM, France
Leon Glass	McGill University, Canada
Pauline Haddow	Norwegian University of Science and Technology, Norway
David Halliday	University of York, UK
Jennifer Hallinan	Newcastle University, UK
Paulien Hogeweg	Utrecht University, The Netherlands
Arun Holden	University of Leeds, UK
Johannes Knabe	University of Hertfordshire, UK
Gareth Leng	University of Edinburgh, UK
Sam Marguerat	University College London, UK
Maizura Mokhtar	University of Central Lancashire, UK

J. Manuel Moreno   Universitat Politècnica de Catalunya, Spain
Chrystopher L. Nehaniv   University of Hertfordshire, UK
Simon O'Keefe   University of York, UK
Hiroshi Okamoto   RIKEN Brain Science Institute, Japan
Tjeerd Olde Scheper   Oxford Brookes University, UK
Heike Sichtig   University of Florida, USA
Leslie Smith   University of Stirling, UK
Nicole Soranzo   Sanger Institute, UK
Denis Thieffry   École Normale Supérieure, France
Christof Teuscher   Portland State University, USA
Jim Tørresen   University of Oslo, Norway
Andy Tyrrell   University of York, UK
Juanma Vaquerizas   European Bionformatics Institute
Simon Whelan   Manchester University, UK

Sponsoring Institutions

Trinity College, Cambridge
Biochemical Journal
Eagle Genomics
Faculty of 1000
Molecular Biosystems

Table of Contents

Genetic and Epigenetic Networks

Transcriptomics and Gene Regulation

Signalling Pathways and Responses

Protein Structure and Metabolic Networks

Patterning and Rhythm Generation

Neural Modelling and Neural Networks

Biomedical Modelling and Signal Processing

Information Processing and Representation

Algorithmic Approaches in Computational Biology

Using Artificial Epigenetic Regulatory Networks to Control Complex Tasks within Chaotic Systems

Alexander P. Turner[1,4], Michael A. Lones[1,4], Luis A. Fuente[1,4],
Susan Stepney[2,4], Leo S. Caves[3,4], and Andy M. Tyrrell[1,4]

[1] Department of Electronics
{apt503,mal503,laf509,amt}@ohm.york.ac.uk
[2] Department of Computer Science
susan.stepney@cs.york.ac.uk
[3] Department of Biology
lsdc1@york.ac.uk
[4] York Centre for Complex Systems Analysis (YCCSA)
University of York, Heslington, York YO10 5DD, UK

Abstract. Artificial gene regulatory networks are computational models which draw inspiration from real world networks of biological gene regulation. Since their inception they have been used to infer knowledge about gene regulation and as methods of computation. These computational models have been shown to possess properties typically found in the biological world such as robustness and self organisation. Recently, it has become apparent that epigenetic mechanisms play an important role in gene regulation. This paper introduces a new model, the Artificial Epigenetic Regulatory Network (AERN) which builds upon existing models by adding an epigenetic control layer. The results demonstrate that the AERNs are more adept at controlling multiple opposing trajectories within Chirikov's standard map, suggesting that AERNs are an interesting area for further investigation.

1 Introduction

Gene regulatory networks are complex structures which underpin an organism's ability to control its internal environment [23]. From a biological perspective, the study of gene regulation is of interest because it determines cellular differentiation which is responsible for the development of the different tissues and organs that underpin the structure of higher organisms [17]. From a computational perspective, gene networks are interesting because they are robust control structures, capable of dealing with serious environmental perturbations, and yet maintaining structure and order. Because of this, there has been significant interest in modeling gene regulatory networks *in silico* in order to capture these features [18,21]. Due to the complexity of biological gene regulation, computational analogues are vastly simplified. However, research has shown that relatively simple networks such as the random Boolean network can exhibit emergent

M.A. Lones et al. (Eds.): IPCAT 2012, LNCS 7223, pp. 1–11, 2012.

properties such as self organisation and robustness and, in addition to this, can model real regulatory circuits [15,14]. It is the balance between complexity and function which is often the limiting factor in the creation of truly analogous computational models.

It is apparent that computational analogues of gene regulation fail to include models of one of the most pervasive methods of gene regulation, epigenetics. In this sense, the regulatory nature of these computational analogues may be limited in terms of complexity and performance. In this paper, we attempt to define a representation of epigenetic information which to some extent captures the useful properties of epigenetics for the control of a complex dynamical system.

2   Gene Regulation and Epigenetics

A gene is a unit of hereditary information within a living organism, most commonly considered to be a section of DNA that specifies the primary structure of a protein. The genetic code is a biological blueprint that details which proteins can be produced, and ultimately, the phenotypic space which the organism can exist within. The lowest known threshold on the number of genes required to naturally facilitate life is that of *Mycoplasma genitalium*, which has approximately 470 predicted genes [9]. Even in nature's most minimalist example of a gene regulatory network, 470 genes have to be coordinated in such a way to maintain the optimum internal environment of the organism, highlighting that even the simplest of gene regulatory networks are inherently complex.

Epigenetic mechanisms allow a further layer of control over gene regulation. This is commonly done by physically restricting the accessibility of the genes. One of the principal epigenetic mechanisms is DNA methylation which refers to the addition of a methyl group to either the cytosine or adenine nucleobase in DNA (Figure 1). It acts as an epigenetic marker which can regulate many physiological processes [3,11]. With increasingly complex organisms, higher order structures such as chromatin have been shown to influence gene expression. Chromatin has two functions; Firstly, in the case of human cells, there are approximately 3400Mb of DNA of approximately 2.3m in length. Chromatin provides a structure for the condensation of DNA, so that it can be fully contained in a nucleus of approximately $6\mu m$ [1,2,6]. Secondly, chromatin modifications have the ability to control access to the DNA, which in turn acts as an additional level of genetic control. Generally, DNA methylation provides a more long term, stable effect on gene expression when compared to relatively short term reversible chromatin modifications [7]. However, research suggests that in some instances chromatin modifications and DNA methylation are intrinsically linked [12].

One of the more interesting aspects of epigenetics is that in certain instances, epigenetic traits can be inherited by successive generations of cells, and sometimes organisms [4]. In addition to this, epigenetic modifications can give the genetic code a relative memory [3], which can then be used in such processes as cellular differentiation [20]. However, the specific mechanisms and processes which control epigenetics, and in turn gene expression, are not yet wholly understood.

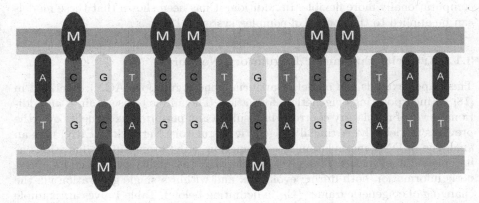

Fig. 1. An illustration of a highly methylated DNA region. The attachment of a methyl group (M) to the cytosine (C) nucleobase (DNA methylation) has been shown to be able to regulate physiological processes [3,11].

Current research demonstrates that epigenetic mechanisms allow for a level of genetic memory, and a method of genetic control above the genetic code itself. This creates the ability for organisms to express high levels of adaptability via utilisation of the extra dynamical processing which epigenetics facilitates. In the following sections the idea of incorporating a level of artificial epigenetic information in pre-existing artificial gene regulatory networks is introduced. Simulations are conducted to ascertain if the addition of epigenetic information can yield any performance increases when controlling two trajectories within Chirikov's standard map.

3   Artificial Genetic Regulatory Networks

Gene regulation in biology is a set of mechanisms which maintains control of an organism's internal environment (homeostasis). The aim of artificial gene regulation is to create a computational model of genetic behaviour which exhibits the useful and interesting properties of gene regulation in nature, namely self organisation and robustness. The first such example was the random Boolean network (RBN) [15]. RBNs represent genes as Boolean expressions. These artificial genes are referred to as node's. The network has a connectivity value (k) which specifies how many node's influence a given nodes expression level. The state of a node is defined by randomly initiated state transition rules. Upon execution, the network is iterated over a number of time steps, during which each node modifies its value depending upon its connectivity and its state transition rules. These networks demonstrate that with a k value of 2 or 3, distinct order and repetitive patterns can be generated. Moreover, for certain parameter ranges, the RBNs expressed high levels of robustness, maintaining relative order when exposed to external perturbations [10].

Subsequent models of gene regulation draw inspiration from the RBN. However there has been a shift towards continuous models [16,18], as they are computationally more flexible. In addition, it has been shown that these models can be applied to the control of complex systems [18].

3.1 Artificial Epigenetic Regulatory Networks

This paper extends the model of artificial gene regulation (AGN) described in [18] by incorporating epigenetic information. The aim is to ascertain if an additional level of regulatory control will result in any performance benefits over the previous model. The Artificial Epigenetic Regulatory Network (AERN) uses an analogue of DNA methylation in combination with chromatin modifications as its epigenetic elements. This gives the network the ability to change its epigenetic information both during evolution, and within a single generation via the changing of epigenetic frames ($\mathbf{E}_G$ in definition below). Table 1 gives an example of an evolved AERN.

The AERN can be formally described as: $< G, L_G, I_G, O_G, E_G >$ where :

  $\mathbf{G}$ = An indexed set of genes $\{g_0, ..., g_n : g_i = < \lambda_i, R_i, f_i >\}$, where:
    $\lambda_i : \mathbb{R}$ is the expression level of a gene
    $R_i \subseteq G$ is the set of regulatory inputs used by the genes
    $f_i : R_i \to \lambda_i$ is a gene's regulatory function
  $\mathbf{L}_G$ is an indexed set of initial expression levels, where, $|L_G| = |G|$
  $\mathbf{I}_G \subset G$ are the external inputs applied to the network
  $\mathbf{O}_G \subset G$ are the outputs of the network
  $\mathbf{E}_G \subset G$ is a data structure specifying which genes are *active* at a given
    instance

The AERN is executed as follows :

  G1. $\lambda_0....\lambda_n$ are initialised from L_G (if AERN not previously executed).
  G2. Expression levels of enzymes in I_G are set by the external inputs.
  G3. At each time step, each *active* gene g_i applies its regulatory function f_i
       to the current expression levels of its *active* regulating genes R_i in order
       to calculate its expression at the next time step, λ_i'
  G4. After a given number of iterations, execution is halted and the
       expression levels of enzymes in O_G are copied to the external outputs.

The expectation is that epigenetic control will enable the network to evolve in such a way that it will have certain genes that are more able to perform a given objective. The inclusion of epigenetic information would give the network the ability to allocate different genes to different tasks, effectively regulating gene expression according to the environment in which it is operating.

Each gene uses a parameterisable sigmoid function which has been shown in [18] to be the most effective in traversing the complex. The parameter settings are summarised in Table 2.

Table 1. Example data attributes for an AERN of size 8. The only difference between the AERNs and the AGNs is the introduction of 2 epigenetic frames, which specify which genes will be active for each objective.

Variable	External Inputs ($\mathbf{I}_G$)		Genes					Outputs ($\mathbf{O}_G{\subset}$G)
Gene Expression Values ($\mathbf{L}_G$)	0.18	0.81	0.54	0.38	0.95	0.14	0.05	0.47
Weights	0.47	-0.27	0.24	0.99	-0.87	-0.02	-0.47	0.97
Sigmoid Offset	-0.18	0.24	0.14	-0.50	-0.21	0.57	0.31	0.38
Sigmoid Slope	1	10	5	19	2	14	3	7
Connections	5	2	1	5	7	3	2	3
	7	4		5	2	7	1	1
	5	2		5		3	2	3
		4		4		1		7
Epigenetic Frame A ($\mathbf{E}_G{\subset}$G)	1	0	1	1	0	0	0	1
Epigenetic Frame B ($\mathbf{E}_G{\subset}$G)	0	1	1	0	1	1	1	1
Network Iterations	15							

4   Dynamical Systems

A dynamical system is a system whose current state is a product of its previous state and an evolution rule. For any given point within a dynamical system, its trajectory through the space is governed by the iteration of this rule over a given time frame. One of the most interesting groups of dynamical systems are those that exhibit chaotic dynamics. Chaos can be recognised as irregular and unpredictable behavior within a system, due to extreme sensitivity to initial conditions [22]. However, unlike random behavior, this is the result of applying deterministic evolution rules.

4.1   Standard Map

Chirikov's standard map [8] is a dynamical system which under certain conditions expresses chaotic dynamics. Its behaviour results from two difference equations :

$$x_{n+1} = (x_n + y_{n+1}) \bmod 1 \qquad y_{n+1} = y_n - \frac{k}{2\pi} \sin(2\pi x_n) \qquad (1)$$

The modification of parameter k has a direct effect on the dynamics of the system (Figure 2). In figures 2a and 2c it can be seen that the dynamics of the map are for the most part homogeneous (with slight differences in the corner of the map). However, when k reaches a critical value of 0.972, there is a distinct increase in the number of elliptic islands, and the distance change at a given co-ordinate is no longer consistent over the majority of the map [22] (Illustrated in figure 2b, where $k = 1$).

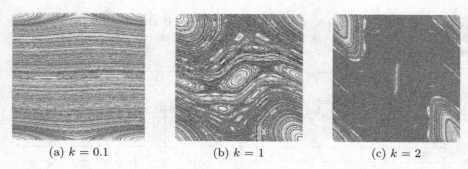

(a) $k = 0.1$ (b) $k = 1$ (c) $k = 2$

Fig. 2. Sample of Chirikov's Standard Map during ordered (a and c) and chaotic (b) dynamical phases

4.2 State Space Targeting

The presence of chaos makes it difficult to predict subsequent points in a trajectory. However, research has shown that targeting within the standard map during periods of chaos is possible via the use of perturbations, allowing navigation to different regions of the state space[18,5].

5 Experimentation

In order to test the relative performance increases of the AERNs, they have been evaluated against the model from [18] when controlling two opposing trajectories within the standard map.

5.1 Evolution of the Networks

A genetic algorithm (GA) is used to evolve the network. GAs are population based search algorithms [19], which often find near optimal solutions within a tractable time frame, which makes it a favorable approach for this experiment. The GA uses a crossover rate of 0.5, a mutation rate of 0.001 and tournament selection of size 4 to evolve networks containing 20 genes. The simulation is run over 50 generations, and the fittest individual at the 50th generation is the score for that run. 50 runs were carried out for each network type.

5.2 State Space Targeting Using Artificial Regulatory Networks

Previous work demonstrates that artificial gene regulatory networks (AGN) can be used to target specific regions of Chirikov's map [18]. This paper builds upon this by using the networks to control two opposing trajectories within the standard map. There were two objectives, which specify that the networks have to guide a trajectory from the bottom to the top of the complex map, and then from the top to the bottom in the lowest number of iterations.

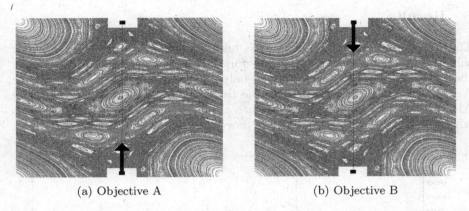

(a) Objective A   (b) Objective B

Fig. 3. Chirikov's standard map showing two start and finish areas for each objective

Each network received 3 inputs and produced 1 output. The three inputs were the x co-ordinate, the y co-ordinate and the distance from the center of the target. The output of the network was the new k value (see Eqn 1), which is used in the next iteration of the equation. Since the output of the network is a real value (Table 2), it must be scaled to the interval [1,1.1] to ensure that the map remains in its chaotic phase. Upon initialisation, the starting point for the trajectory is a randomly initiated point within the starting region, and the networks are randomly initiated with values shown from Table 2. The epigenetic frame changes when either the first objective is completed, or the the maximum number of steps for that objective has been reached.

For the objective specifying the trajectory is to travel from the bottom to the top of the map, the initial start point is [0,0.05] for the y co-ordinate, and [0.45,0.55] for the x co-ordinate. For the second objective, the randomly initialised start point is between [0.95,1] for the y co-ordinate, and [0.45,0.55] for the x co-ordinate. The score for that run is the average number of steps to traverse the map in both directions, up to a maximum number of 1000 steps. During execu-

Table 2. Parameters of the variables within the AERN

Variable	Type	Range
Gene Expression	Real	0;1
Weights	Real	-1;1
Sigmoid Offset	Real	-1;1
Sigmoid Slope	Int	0;20
Epigenetic Objective	Int	0;1
Network Iterations	Int	1;20

tion, the average of the 40 runs (20 for each objective) are collated together, and if there is an instance where the network completed both objectives, the fitness is rewarded by 250 steps. If only one objective is completed, the score is decremented by 100 steps up to a maximum of 1000 steps (the final results will not take into consideration rewards and punishments, and only solutions that complete both objectives are able to achieve a score less than 1000). This is to place emphasis on the networks completing both objectives, whilst allowing strong solutions that only complete one objective to remain in the population.

6   Results

The results show that both the AGNs and the AERNs were able to produce solutions that could control both trajectories. However, the results indicate that the tasks were difficult (Figure 4), as only 35% of the AGN's and 52% of the AERNs were able to complete both objectives. The results demonstrate that the AERNs perform significantly better than the AGNs. The trend is more evident when the unsuccessful instances are removed, as can be seen in Figure 5.

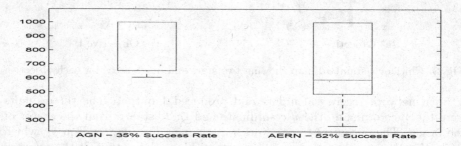

Fig. 4. Summary of the fitness distributions (low numbers are better). The results for the AGN show less than a 50% success rate, and a best value of approximately 600 steps. The AERNs show a significant increase in success rate, and better solutions overall.

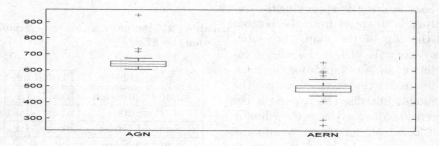

Fig. 5. Data from Figure 4 with the unsuccessful results omitted (low numbers are better). This shows a much clearer trend in the data, and the addition of the artificial epigenetic information can be seen to increase performance by approximately 150 steps on average.

One of the more interesting aspects of the results is that the data in Figure 5 shows that the vast majority of the successful AERNs outperform almost all of the successful results from the AGNs, and yet the trends of the data are quite similar. On average the AERNs produce an improvement of approximately 150 steps in terms of mean path length.

Aside from increased performance, the AERNs demonstrate further benefits. First, with the ability to inactivate genes comes the ability to increase the efficiency of the networks. Hence, with each inactive gene for each objective, there is less computational effort required to complete a single iteration of the network. Over the entirety of a simulation, this could lead to significant performance increases. A further advantage is that the epigenetic layer of the networks provides a source of qualitative analysis. By dissection of the network structures, it could be seen that certain input variables are not needed in the navigation of the complex map. In some of the best solutions, the input variable for the x co-ordinate was not required, which provides additional information about the problem space that otherwise would not have been readily available.

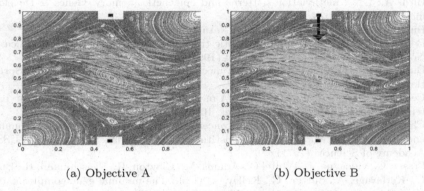

(a) Objective A   (b) Objective B

Fig. 6. Example behaviour of the solutions produced by the AERNs. Objective A was completed in 113 steps, and objective B was completed in 329 steps.

7 Conclusions

This paper has illustrated the potential for incorporating epigenetic information in computational models of gene regulation, and the initial results are highly promising. The results demonstrate that evolved AERNs are able to assign certain genes to certain tasks, improving functionality and efficiency. This ties in well with the biology of epigenetics, which allow for a higher level of genetic control without compromising efficiency [13].

There is a significant amount of further research required to assess the full functionality of the AERNs. In future work, the AERNs will be applied to a range of tasks to best evaluate their strengths. Additionally, the topologies of the networks will be looked at in more detail to ascertain the role and function of varying epigenetic mechanisms. Furthermore, research will be conducted into introducing a metabolic network to attempt to best utilise the efficiency potential of the AERNs.

Acknowledgments. This work is supported by the EPSRC funded project (ref:EP/F060041/1), Artificial Biochemical Networks: Computational Models and Architectures.

Thank you to the reviewers for their comments and feedback.

References

1. Alberts, B., Bray, D., Lewis, J., Raff, M., Roberts, K., Watson, J.: Molecular Biology of the Cell, 3rd edn. Oxford Univ. Press (1994)
2. Allis, C., Jenuwein, T., Reinberg, D.: Epigenetics. Cold Spring Harbor Laboratory Press (2007)
3. Bird, A.: DNA methylation patterns and epigenetic memory. Genes & Development 16(1), 6 (2002)
4. Bird, A.: Perceptions of epigenetics. Nature 447(7143), 396 (2007)
5. Bollt, E., Meiss, J.: Controlling chaotic transport through recurrence. Physica D: Nonlinear Phenomena 81(3), 280–294 (1995)
6. Bushman, F.: Lateral DNA transfer: mechanisms and consequences. Cold Spring Harbor Laboratory Press (2002)
7. Cedar, H., Bergman, Y.: Linking DNA methylation and histone modification: patterns and paradigms. Nat. Rev. Genet. 10(5), 295–304 (2009)
8. Chirikov, B., Sanders, A.: Research concerning the theory of non-linear resonance and stochasticity. Nuclear Physics Institute of the Siberian Section of the USSR Academy of Sciences (1971)
9. Fraser, C., Gocayne, J., White, O., Adams, M., Clayton, R., Fleischmann, R., Bult, C., Kerlavage, A., Sutton, G., Kelley, J., et al.: The minimal gene complement of mycoplasma genitalium. Science 270(5235), 397 (1995)
10. Harvey, I., Bossomaier, T.: Time out of joint: Attractors in asynchronous random boolean networks. In: Proceedings of the Fourth European Conference on Artificial Life, pp. 67–75. MIT Press, Cambridge (1997)
11. Hattman, S.: Dna-[adenine] methylation in lower eukaryotes. Biochemistry 70(5), 550–558 (2005)
12. Jackson, J., Lindroth, A., Cao, X., Jacobsen, S.: Control of CpNpG DNA methylation by the kryptonite histone h3 methyltransferase. Nature 416(6880), 556–560 (2002)
13. Jeanteur, P.: Epigenetics and Chromatin. Progress in Molecular and Subcellular Biology. Springer, Heidelberg (2008)
14. Kauffman, S., Peterson, C., Samuelsson, B., Troein, C.: Random Boolean network models and the yeast transcriptional network. Proceedings of the National Academy of Sciences of the United States of America 100(25), 14796–14799 (2003)
15. Kauffman, S.: Metabolic stability and epigenesis in randomly constructed genetic nets. Journal of Theoretical Biology 22(3), 437–467 (1969)
16. Kumar, S.: The evolution of genetic regulatory networks for single and multicellular development. In: GECCO. Citeseer (2004)
17. Latchman, D.S.: Gene regulation: a eukaryotic perspective. Advanced text. Taylor & Francis (2005)
18. Lones, M.A., Tyrrell, A.M., Stepney, S., Caves, L.S.: Controlling Complex Dynamics with Artificial Biochemical Networks. In: Esparcia-Alcázar, A.I., Ekárt, A., Silva, S., Dignum, S., Uyar, A.Ş. (eds.) EuroGP 2010. LNCS, vol. 6021, pp. 159–170. Springer, Heidelberg (2010)

19. Mitchell, M.: An introduction to genetic algorithms. Complex adaptive systems. MIT Press (1998)
20. Mohn, F., Schübeler, D.: Genetics and epigenetics: stability and plasticity during cellular differentiation. Trends in Genetics 25(3), 129–136 (2009)
21. Quick, T., Nehaniv, C.L., Dautenhahn, K., Roberts, G.: Evolving Embodied Genetic Regulatory Network-Driven Control Systems. In: Banzhaf, W., Ziegler, J., Christaller, T., Dittrich, P., Kim, J.T. (eds.) ECAL 2003. LNCS (LNAI), vol. 2801, pp. 266–277. Springer, Heidelberg (2003)
22. Tél, T., Gruiz, M.: Chaotic dynamics: an introduction based on classical mechanics. Cambridge University Press (2006)
23. Turner, B.: Chromatin and gene regulation: mechanisms in epigenetics. Blackwell Science (2001)

A Gene Regulatory Network Simulation of Heterosis

Peter Martin Ferdinand Emmrich, Hannah Elizabeth Roberts, and Vera Pancaldi

Department of Plant Sciences, University of Cambridge
{pmfe2,hr283,vp301}@cam.ac.uk

Abstract. We describe a simulation of multi-genic heterosis using Boolean gene regulatory networks. Hybrid vigor, or heterosis, is the phenomenon whereby the offspring of crosses from separate populations often perform better than inbreds with respect to growth rate, fertility and disease resistance. Because of its great economic importance, the genetic and molecular basis of heterosis has been subject of many scientific studies. However, attempts to model the phenomenon from a systems biology point of view have been quite abstract.

Our model allows the generation, evolution, homologous recombination and hybridisation of networks which display the properties of gene regulatory networks observed in biology in a simulated environment. We can thus test the current hypotheses about heterosis and investigate which factors affect it.

Keywords: Boolean models, biological networks, R , BoolNet, heterosis, hybrid vigor, evolutionary algorithm.

1   Introduction

1.1   Biological Background

Hybrid vigor, or heterosis, is the phenomenon whereby the offspring of crosses from separate populations often perform better than inbreds with respect to growth rate, fertility and disease resistance. It has been observed across biological kingdoms and the effects have been known about for many years. For example, Darwin observed that crossing unrelated parents is useful in plant and animal breeding and that wild plants and animals have evolved means of favoring cross-fertilization over inbreeding. Today, hybrid vigor is being exploited to a great extent in agriculture and animal breeding [1].

Because of its great economic importance, the genetic and molecular basis of heterosis has been subject of many scientific studies. Three non-exclusive models have been proposed for the emergence of single-trait heterosis: dominance, overdominance and pseudo-overdominance. According to the dominance model, the hybrid is vigorous because the alleles causing the improved phenotype are dominant. The overdominance model claims that the heterozygosity is itself advantageous, as it confers a greater molecular and regulatory diversity to the organism. The pseudo-overdominance model is an extension of the dominance model, which takes gene linkage into account. Dominant alleles that segregate with recessive alleles for other genes may give the appearance of overdominance. However, these models have been

M.A. Lones et al. (Eds.): IPCAT 2012, LNCS 7223, pp. 12–16, 2012.

criticized for inadequately explaining heterosis in multi-genic traits [2]. A systems biology approach might provide valuable insight into the causes of hybrid vigor. Previous studies in this direction have been highly abstract [3],[4].

We used biologically realistic simulated evolution of diploid populations to build a more comprehensive model and test several hypotheses about hybrid vigor:

1. Levels of heterosis increase with genetic distance of the parents up to a maximum, but decrease sharply after that due to hybrid incompatibility.
2. Dominance, overdominance and pseudo-overdominance are insufficient as explanatory models for hybrid vigor.
3. Hybrid networks display higher connectivity than the inbred parent networks.
4. The type of selection pressure on the population (for example whether fitness is defined based on a constant or varying environment) determines the extent of hybrid vigor.

1.2   Random Boolean Networks

Since the first papers on the subject [5], Boolean logic has been used to model genetic regulatory networks [6].The assumption that genes can only be ON or OFF, with no intermediate values, allows us to simulate large networks relatively easily and to represent the state of the network as a combination of states of single nodes. At the same time the different biological mechanisms involved in gene regulation can readily be represented by logic functions. Once the connections between the nodes in a Boolean network are specified, the attractors of the system can be found. These are states or periodic succession of states to which many different initial conditions lead. We adopt this framework to model gene regulatory networks including their interaction with a simulated environment. Nodes represent genes/proteins and directed edges represent regulatory interactions of any kind.

2   Network Construction

2.1   Generating Networks with Realistic Properties

Several properties have been described as characteristic for biological networks. Such networks have been mapped in several model organisms, showing that specific statistical network properties are conserved across species and also suggesting mechanisms for how evolution might influence the network properties. To achieve a realistic network topology, we used the algorithm described by Holme and Kim [7], which produces scale-free networks with the small-world property and realistic clustering. We adapted the algorithm to generate directed networks, which we use to specify the regulatory connections between genes in our simulated individuals. Once the network is generated we assign a random initial transition function to every node. This function describes the binary state that a node will be set to, depending on all of its ingoing connections. Thus the transition function represents the logical wiring of the inputs into the node.

2.2   Selection of Phenotypes Based on Network Responses to Several Environments

Once these networks are generated, we use the R package BoolNet [8] to handle the network and simulate Boolean dynamics on it. At the point of network generation, a number of nodes are specified as environment nodes. These do not have any ingoing connections and, by setting them on or off in different combinations, environments for the simulations can be generated with every combination representing one 'environment'. The interaction between the simulated population and the environment is implemented through the association of environments to sub-networks, akin to pathways in a biological context. An algorithm selects modules within the network, which are characterised by high clustering coefficients, and each environment is assigned one of these output modules. For each environment, fitness is defined as the extent to which the modules activity reflects the pattern required by the specific environment. Selection thus drives the network towards turning on the nodes of the associated output module, while turning off the nodes of other output modules.

The combined phenotype of each network is calculated from the phenotypes it exhibits under every environment: For each environment, the attractors of the network are found by a heuristic method. Attractor phenotypes are calculated by averaging the states of nodes in the output module over the number of involved states and subtracting the average of other output module states. This average can be weighted to allow for different environmental emphases, for example making a difference between fitness based on performance in multiple environments or based on performance in a single environment. The phenotype under one environment is calculated by averaging the phenotypes of every attractor, using a weighted average, depending on the basin sizes of the attractors (number of initial states that lead into each).

3   Network Evolution and Hybrid Vigor

3.1   Evolution Is Achieved by Mutation and Selection

To allow networks to evolve, we developed a number of functions to introduce mutations. These include the loss of an ingoing or outgoing connection of a node, addition of a connection between two nodes, node loss and node gain by duplication [9]. We studied how the relative rates of different mutations affect the degree distribution, clustering coefficient and average path length of the network over the course of evolution. For the further experiments, we adjusted the mutation rates to preserve realistic network properties.

Our model simulates evolution by iterating several cycles of mutation and selection. Each generation is formed by replicating and mutating the individuals of the previous generation which display the best network phenotypes. Individuals can be simulated as haploids or diploids. In the case of diploid evolution, the two networks are superimposed and appropriate cross-connections are added between them. If a node has ingoing connections from nodes of both networks, these are linked via an OR-gate. In biological terms, this reflects dominant-recessive relationships. In the diploid model, calculation of phenotypes and selection occurs at the stage of diploids, while mutations are introduced during the haploid stage. We reduce the computational complexity of the phenotype calculation by considering node pairs with identical connections as single (homozygous) nodes.

Haploid gametes are generated by picking one node from every pair of haploid parents' nodes at random. Then gametes are combined to make diploid offspring. The concept of genetic linkage, where alleles do not segregate independently, is simply implemented by assigning nodes to groups and imposing inheritance at the group level rather than at the single node level.

3.2   Haploid and Diploid Networks Evolve towards Better Phenotypes

Simulations for haploid and diploid evolution were run for a population of one hundred nodes for one hundred generations and with ten offspring per pair of parents. In the haploid model we observe a sharp increase in combined phenotypes at the start of the simulation. After many generations any further increases are marginal. Increases in phenotypes often occur in bursts rather than by slow shift. The speed of evolution is strongly dependent on several factors including network size, rates of different mutations and population size. Without homologous recombination, the diploid phenotype function initially behaves like the haploid, but then enters a phase of decline. This effect is more pronounced in small populations due to high rates of inbreeding. Introducing homologous recombination mixes the genetic material and allows us to explore the various mechanisms that underlie hybrid vigor.

3.3   Simulating Hybrid Vigor

We generated an initial haploid network, which was replicated with mutations and combined to form an initial population of diploids. We allow this population to adapt to a set of environments by means of diploid evolution with homologous recombination (A in Fig. 1). From the adapted population we select the homozygote with the highest combined phenotype to be the common ancestor of the different lines. Its offspring is separated into several subpopulations after a further stage of diploid evolution (B). Diploid formation and homologous recombination are then only allowed within a subpopulation (C). At regular intervals we compare the phenotypes of crosses between subpopulations (hybrids) to intra-population crosses (D). This allows us to measure the effects of hybridization on the phenotype as a function of the genetic distance between the parents.

Our preliminary results show that this model simulates hybrid vigor, shown by consistently higher phenotypes of the cross-population hybrids as compared to intra-population diploids.

Repeating this on multiple populations and changing some of the parameters allows us to estimate the impact on hybrid vigor of the different selection and mutation mechanisms.

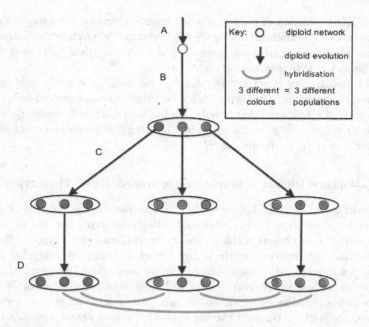

Fig. 1. subpopulations evolve separately, the phenotypes of hybrids are compared to crosses within subpopulations

References

1. Goff, S.A.: A unifying theory for general multigenic heterosis: energy efficiency, protein metabolism, and implications for molecular breeding. New Phytol. 189, 923–937 (2011)
2. Birchler, J.A., Yao, H., Chudalayandi, S., Vaiman, D., Veitia, R.A.: Heterosis. The Plant Cell 22, 2105–2112 (2010)
3. Brummitt, C.: Network Model of Heterosis ECS289L Project Report (2011)
4. Andorf, S.: Molecular network structures in heterozygotes: A systems-biology approach to heterosis. J. Bioinform. Syst. Biol. (2009)
5. Kauffman, S.A.: Metabolic stability and epigenesis in randomly constructed genetic nets. J. Theor. Biol. 22, 437–467 (1969)
6. Schlitt, T., Brazma, A.: Current approaches to gene regulatory network modelling. BMC Bioinformatics 8, S9 (2007)
7. Holme, P., Kim, B.J.: Growing scale-free networks with tunable clustering. Phys. Rev. E Stat. Nonlin. Soft Matter Phys. 65, 026107 (2002)
8. Müssel, C., Hopfensitz, M., Kestler, H.A.: BoolNet–an R package for generation, reconstruction and analysis of Boolean networks. Bioinformatics 26, 1378–1380 (2010)
9. Teichmann, S.A., Babu, M.M.: Gene regulatory network growth by duplication. Nature Genetics 36, 492–496 (2004)

Comparing Discrete and Piecewise Affine Differential Equation Models of Gene Regulatory Networks

Shahrad Jamshidi, Heike Siebert, and Alexander Bockmayr

Freie Universität Berlin, DFG Research Center Matheon
Arnimallee 6, D-14195 Berlin, Germany
{jamshidi,hsiebert,bockmayr}@zedat.fu-berlin.de

Abstract. We compare the discrete asynchronous logical modeling formalism for gene regulatory networks due to R. Thomas with piecewise-affine differential equation models. We show that although the two approaches are based on equivalent information, the resulting qualitative dynamics are different.

Keywords: Gene regulatory networks, mathematical modeling, discrete models, piecewise-affine models.

1 Introduction

Gene regulation is the result of the complex interplay of molecular components forming large interaction networks. Mathematical modeling of gene regulatory networks gives insights into the underlying structure and dynamics of various biological systems. If information on kinetic parameters is lacking, *qualitative formalisms* offer a well-established alternative to the more traditionally used differential equation models. Using only qualitative information on the network structure and the interactions between the components, these approaches allow obtaining an abstract description of the system's dynamics.

The discrete formalism of R. Thomas [1] is a qualitative method describing a gene regulatory network by a discrete function. Each network component is represented by a variable that takes integer values representing the different levels of gene activity. The information on how the behavior of one component is governed by the values of the other components is captured in a discrete function. The component functions then constitute the coordinate functions of the update function of the network. To derive the dynamics of the system, Thomas introduced the asynchronous update method where only one variable changes per discrete time step, and only by a unit value. Since the state space is finite, the dynamics can be represented by a directed graph, the so-called asynchronous *state transition graph* (STG).

The particularities of the asynchronous update method result in a close correspondence of the discrete model to certain differential equation systems [2].

M.A. Lones et al. (Eds.): IPCAT 2012, LNCS 7223, pp. 17–24, 2012.

Differential equation models using step functions have a continuous time evolution, yet can be seen as qualitative due to the close relation of step and discrete functions. Such piecewise affine differential equation (PADE) models approximate certain ordinary differential equation models [3,4]. De Jong et al. [5] have shown that they can essentially be captured by a discrete representation which abstracts the continuous solution trajectories of the differential equations into transitions between different regions of the phase space. Again, the resulting dynamics can be represented by a digraph, the *qualitative transition graph* (QTG).

In this paper, we aim at clarifying the relation between Thomas and PADE models by comparing the respective graphs capturing the dynamical behavior. Several results in this direction already exist. For example, attractors, including steady states and certain limit cycles, are related [2,3,6–8]. Our goal here is to present a comprehensive comparison between the STG and the QTG.

The paper is organized as follows. Sect. 2 presents a discrete modeling approach based on the Thomas formalism. PADE systems and the qualitative analysis developed by de Jong et al. [5] are introduced in Sect. 3. In Sect. 4, we show that the two approaches use equivalent information. Sect. 5 contains our main result characterizing transitions in the QTG using edges originating in corresponding vertices in the STG. We illustrate the application of this result with examples of relations between paths and attractors in the two graphs. The conclusion and perspectives for future work are given in Sect. 6.

This paper is an extended abstract intended to illustrate more formal results.

2   Discrete Formalism

Consider a gene network with n regulatory components. In the discrete modeling approach, the activity level of a component i is modeled by a discrete variable q_i, which takes its values in a finite set of natural numbers $Q_i = \{0, \ldots, p_i\}$. The *state space* of the discrete model is $Q = Q_1 \times \cdots \times Q_n$, and the regulatory interactions are captured by a discrete *update function* $f = (f_1, \ldots, f_n) : Q \to Q$. The function f uniquely determines the *state transition graph* $STG(f) = (Q, E)$, a directed graph with node set Q and edge set $E \subset Q \times Q$. For any $j \in \{1, \ldots, n\}$, $q \in Q$ with $f_j(q) \neq q_j$, there is an edge $(q, q') \in E$, where $q'_j = q_j + sgn(f_j(q) - q_j)$ and $q'_i = q_i$, for all $i \in \{1, \ldots, n\} \setminus \{j\}$. Here, $sgn : \mathbb{R} \to \{-1, 0, 1\}$ denotes the sign function. If $f(q) = q$, then $(q, q) \in E$ and q is called a *fixpoint*.

Example 1. For $Q = \{0, 1\} \times \{0, 1, 2\}$ and the update function $f : Q \to Q$

$$\frac{q \quad |00|01|02|10|11|12}{f(q)|12|12|11|00|10|11}$$

the state transition graph $STG(f)$ is displayed on the left of Fig. 1(a).

3   Piecewise Affine Differential Equations

Next we discuss piecewise affine differential equations (PADE) and the qualitative modeling approach introduced by de Jong et al. [5].

Consider an n-dimensional phase space $\Omega = \Omega_1 \times \cdots \times \Omega_n \subset \mathbb{R}^n_{\geq 0}$, where $\Omega_i = \{x_i \in \mathbb{R} \mid 0 \leq x_i \leq \max_i\}$, and $\max_i \in \mathbb{R}_{>0}$. For every continuous variable $x_i \in \Omega_i$ we assume $p_i \in \mathbb{N}$ thresholds $\theta_i^1, \cdots, \theta_i^{p_i}$ satisfying the ordering

$$0 < \theta_i^1 < \cdots < \theta_i^{p_i} < \max_i, \quad \text{for all } i \in \{1, \ldots, n\}. \tag{1}$$

In the comparison with the discrete formalism in Sect. 2, the value p_i chosen here corresponds to the maximal value p_i of the component range Q_i of a discrete model. We consider a set of PADEs in Ω of the form

$$\dot{x}_i = F_i(x) - G_i(x)x_i, \quad i \in \{1, \ldots, n\}, \tag{2}$$

where the functions $G_i : \Omega \to \mathbb{R}_{>0}$ and $F_i : \Omega \to \mathbb{R}_{\geq 0}$ are linear combinations of products of step functions $S^+(x_l, \theta_l^k) = \begin{cases} 0 & \text{if } x_l < \theta_l^k, \\ 1 & \text{if } x_l > \theta_l^k, \end{cases}$ and $S^-(x_l, \theta_l^k) = 1 - S^+(x_l, \theta_l^k)$ for $l \in \{1, \ldots, n\}$.

To obtain a discrete representation of the PADE system, the state space is partitioned into a set of domains.

Definition 1. *Consider a set of PADEs of the form (2) with phase space Ω and thresholds θ_i^j. The $(n-1)$-dimensional hyperplanes corresponding to the equations $x_i = \theta_i^j$, $j \in \{1, \ldots, p_i\}$, divide Ω into hyper-rectangular regions called domains. A domain $D \subset \Omega$ is defined by $D = D_1 \times \cdots \times D_n$ where every D_i is given by one of the following equations*

$$D_i = \{x_i \mid 0 \leq x_i < \theta_i^1\},$$
$$D_i = \{x_i \mid \theta_i^k < x_i < \theta_i^{k+1}\} \text{ for } k \in \{1, \ldots, p_i - 1\},$$
$$D_i = \{x_i \mid \theta_i^{p_i} < x_i \leq \max_i\},$$
$$D_i = \{x_i \mid x_i = \theta_i^k\} \text{ for } k \in \{1, \ldots, p_i\}.$$

By $\mathcal{D}$ we denote the set of all domains in Ω. A domain $D \in \mathcal{D}$ is called a singular domain, *if there exists $i \in \{1, \ldots, n\}$ such that $D_i = \{x_i \mid x_i = \theta_i^k\}$ for some $k \in \{1, \ldots, p_i\}$. The variable x_i is then called* singular variable. *The order of a singular domain is the number of its singular variables. A domain $D \in \mathcal{D}$ is called a* regular domain, *if it is not a singular domain. The set of regular and singular domains are denoted by $\mathcal{D}_r$ and $\mathcal{D}_s$ respectively.*

It follows immediately that for any regular domain $D \in \mathcal{D}_r$, the functions $F_i(x)$ and $G_i(x)$ are constant on D. Thus (2) can be written as a linear system $\dot{x} = F^D - G^D x$, for all $x \in D$, where $G^D = diag(G_1^D, \ldots, G_n^D)$ is a diagonal matrix with strictly positive entries and $F^D = (F_1^D, \ldots, F_n^D)$ a positive vector. It is easy to see that solutions of (2) starting in a regular domain D converge monotonically towards the so-called *focal point* $\phi(D) := (G^D)^{-1}F^D$.

In agreement with [5], we assume that all focal points lie in a regular domain. By definition of the regular domains, we can then encode the position of each focal point by strict inequalities using the threshold values, and thus obtain a parameter constraint of the form (1) consisting of threshold values and components of the focal point. We call these constraints *ordering constraints*.

To define a suitable dynamics of (2) on singular domains, the differential equations are extended to differential inclusions, and methods presented in [5, 9, 10], give us so-called Fillipov solutions of the differential inclusions. However, our focus here is on the qualitative dynamics, which does not depend on the particularities of the Fillipov extension.

It is shown in [5] that we can calculate the qualitative dynamics of a PADE system (2) using only the respective ordering constraints. This dynamics is represented by a directed graph, the *qualitative transition graph* $QTG(\mathcal{A}) = (\mathcal{D}, \mathcal{T})$. Here, the node set $\mathcal{D}$ consists of all regular and singular domains, and the arcs indicate the existence of suitable solution trajectories of (2) between adjacent domains. As shown in [5], all systems in the class of PADE systems satisfying the same ordering constraints have the *same* qualitative dynamics, i.e., isomorphic QTGs.

Example 2. Consider the system of PADEs

$$\dot{x}_1 = \alpha_1 + \beta_1 S^+(x_1, \theta_1^1) S^-(x_2, \theta_2^1) - \lambda_1 x_1,$$
$$\dot{x}_2 = \alpha_2 + \beta_2 S^+(x_1, \theta_1^1) S^-(x_2, \theta_2^2) + \gamma_2 S^-(x_1, \theta_1^1) S^-(x_2, \theta_2^2) - \lambda_2 x_2.$$

The system has six regular domains with corresponding focal points, e.g., the focal point of $D = [0, \theta_1^1) \times [0, \theta_2^1)$ being $(\frac{\alpha_1}{\lambda_1}, \frac{\alpha_2 + \gamma_2}{\lambda_2})$. We impose the ordering constraints $0 < \frac{\alpha_1 + \beta_1}{\lambda_1} < \theta_1^1 < \frac{\alpha_1}{\lambda_1} < \max_1$ and $0 < \frac{\alpha_2 + \beta_2}{\lambda_2} < \theta_2^1 < \frac{\alpha_2}{\lambda_2} < \theta_2^2 < \frac{\alpha_2 + \gamma_2}{\lambda_2} < \max_2$. The resulting QTG is given on the right of Fig. 1(a).

4   Relating the Discrete and the PADE Formalism

Now we show that the PADE and the discrete formalism contain the same information in the sense that we can transform a PADE system with given ordering constraints into a discrete update function and vice versa.

To obtain a discrete update function from a PADE system we can use a straightforward method originally proposed by Snoussi [2]. First, we discretize the continuous phase space of the PADE system according to its threshold values.

Definition 2. *Let $\mathcal{A}$ be a set of PADEs as in (2), where each variable x_i has p_i ordered threshold values. Let $Q := Q_1 \times \cdots \times Q_n$, where $Q_i := \{0, 1, \ldots, p_i\}$, $i \in \{1, \ldots, n\}$. Define the bijective mapping $d^{\mathcal{A}} : \mathcal{D}_r \to Q$, where*

$$d_i^{\mathcal{A}}(D) := \begin{cases} 0 & \text{if } D_i = \{x \in \mathbb{R} \mid 0 \leq x < \theta_i^1\}, \\ q & \text{if } D_i = \{x \in \mathbb{R} \mid \theta_i^q < x < \theta_i^{q+1}\}, \\ p_i & \text{if } D_i = \{x \in \mathbb{R} \mid \theta_i^{p_i} < x \leq \max_i\}. \end{cases}$$

Second, we exploit the localization of the focal points in the regular domains in order to construct an update function $f^{\mathcal{A}} : Q \to Q$ on the discretized state space Q that shares the dynamical properties of the PADE system $\mathcal{A}$ (see [2] for details). The function $f^{\mathcal{A}}$ is uniquely determined by the ordering constraints for $\mathcal{A}$.

Consequently, the set of PADE systems $\mathcal{A}$ satisfying given ordering constraints can be associated with a single discrete update function $f^{\mathcal{A}}$.

Conversely, a discrete update function can easily be transformed into a PADE system that shares the qualitative dynamical properties.

Definition 3. *Let $f : Q \to Q$ be an update function. Define the discretization function d corresponding to the thresholds $\theta_j^k = k - \frac{1}{2}$ for $j \in \{1, \ldots, n\}$, $k \in \{1, \ldots, p_j\}$, according to Def. 2. We denote by $PADE(f)$ the system of PADEs on $\Omega := \prod_{i=1}^n [0, \max_i]$, $\max_i \in \mathbb{R}_{>p_i}$ for all $i \in \{1, \ldots, n\}$, of the form*

$$\dot{x} = F(x) - x_i, \quad where \quad F(x) = \sum_{q \in Q} f(q) \prod_{j=1}^n S(x, q)$$

and S is composed componentwise of products of step functions yielding $S(x, q) = 1$ if $x \in d^{-1}(q)$, and $S(x, q) = 0$ otherwise.

The choice of threshold values is generic, ensuring an obvious correspondence between the values $0, 1, \ldots, p_i$ in Q_i, $i \in \{1, \ldots, n\}$, and the intervals $[0, \theta_i^1)$, $(\theta_i^k, \theta_i^{k+1})$ for $k \in \{1, \ldots, p_i - 1\}$, and $(\theta_i^{p_i}, \max_i]$. If we calculate the regular domains according to the threshold values and their focal points, we have $\phi(D) = F(x) = f(d(D))$ for all $x \in D$, where $d := d^{PADE(f)}$ and $D \in \mathcal{D}_r$. Equivalently, it holds that $\phi(d^{-1}(q)) \in d^{-1}(f(q))$ for $q \in Q$, which immediately implies that the focal points of $PADE(f)$ satisfy the set of corresponding ordering constraints.

Using these two transformations, we can associate a class of PADE systems characterized by their ordering constraints with a unique discrete update function, and vice versa. The information necessary for constructing the STG resp. QTG is inherent in both representations. In that sense, we can identify every STG with a QTG and vice versa. In fact, it can easily be shown that this identification can be achieved solely on the level of the graphs representing the respective dynamics, although the graphs may hold less specific information than the corresponding discrete functions resp. PADE systems. That is, already the information inherent in the STG is enough to construct the corresponding QTG and vice versa.

Example 3. The function f from Ex. 1 generates $PADE(f)$ whose parameter values satisfy the ordering constraints of the PADE system $\mathcal{A}$ from Ex. 2. Thus $PADE(f)$ belongs to the PADE class represented by $\mathcal{A}$. Similarly, if we discretize $\mathcal{A}$ using Snoussi's method, we obtain the update function f from Ex. 1.

5   Comparing the Dynamics

Although the STG and QTG can be obtained from each other, the qualitative dynamics represented by the two graphs is not the same. Next we analyze differences and similarities of $STG(f)$ and $QTG(\mathcal{A})$ for a discrete update function $f : Q \to Q$ and the corresponding PADE system $\mathcal{A} := PADE(f)$.

Initially, we compare the node sets. The discretization in Sect. 4 implies that the vertices of $STG(f)$ correspond to the regular domain vertices of $QTG(\mathcal{A})$.

· However, there is no representation of the singular domains in the purely discrete setting. To overcome this problem we associate with every singular domain D the set $H(D) \subset Q$ corresponding to the discretization of those regular domains D' that have D in their *boundary* $\partial D'$ (cf. [5]). We thus introduce the mapping

$$H(D) := \begin{cases} \{d(D)\}, & \text{if } D \in \mathcal{D}_r, \\ \{d(D') \in Q \mid D \subset \partial D', D' \in \mathcal{D}_r\} & \text{if } D \in \mathcal{D}_s. \end{cases}$$

Now we are able to state our main result on the correspondences between edges in $QTG(\mathcal{A}) = (\mathcal{D}, \mathcal{T})$ and $STG(f) = (Q, E)$.

Theorem 1. *Let $D \in \mathcal{D}$ and $D' \subset \partial D$. Denote by I the index set of singular variables in D and I' the index set of singular variables in D'. Then*

1. *$(D, D') \in \mathcal{T}$ if and only if*

    (a) *for all $i \in I$ there exist $q^1, q^2 \in H(D)$, $q^1 \neq q^2$, such that $p_i^1 \leq q_i^1$, $q_i^2 \leq p_i^2$ for all $p^1, p^2 \in Q$ with $(q^1, p^1), (q^2, p^2) \in E$, and*

    $$\exists\, l \in \{1, 2\},\, p \in H(D):\ (q^l, p) \in E\ \wedge\ p_i \neq q_i^l,\ \text{if } q_i^1 = q_i^2,$$
    $$\exists\, p^1, p^2 \in H(D):\ (q^1, p^1), (q^2, p^2) \in E\ \wedge\ p_i^1 < p_i^2,\ \text{if } q_i^1 > q_i^2,$$

    (b) *for all $i \in I' \setminus I$ there exists $q \in H(D)$ and $q' \in H(D') \setminus H(D)$ with $q_i \neq q_i'$ and $(q, q') \in E$,*

2. *$(D', D) \in \mathcal{T}$ if and only if*

    (a) *condition 1.(a) holds, and*

    (b) *for all $i \in I' \setminus I$ there exists $q \in H(D)$ and $q' \in H(D') \setminus H(D)$ such that $q_i \neq q_i'$, $q_j' = q_j$ for all $j \neq i$ and $(q, q') \notin E$.*

This result provides the basis for elucidating the correspondences between more complex structures, such as paths or attractors. On the one hand, it can be used for proofs building on local considerations concerning the edges involved and thus confirming some previous findings [7, 10]. On the other hand, it provides ideas for the construction of counterexamples, some of which we present here to illustrate that the relation between the two dynamics is not clear-cut.

We start by considering reachability properties. In simple cases, we can find conditions ensuring the existence of corresponding paths. However, reachability properties are not conserved in general, as can be seen from Fig. 1 (a). There, state $(0,2)$ is reachable from $(1,0)$ in the STG via the path indicated in gray. In the QTG, all paths starting in the regular domain corresponding to $(1,0)$ and all adjacent singular domains do not cross the first threshold plane of the second component. In Fig. 1 (b), we see by considering paths from $(0,0)$ to $(1,1)$ that reachability properties of the QTG are also not conserved in the STG.

Similarly, we are able to find correspondences of attractors (terminal strongly connected components of the directed graphs) including certain steady states, i.e., singleton attractors, and limit cycles, in accordance with [2,3,6–8]. However, the situation becomes less clear if we consider more general attractors.

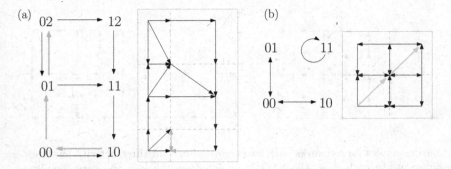

Fig. 1. Corresponding STGs and QTGs. The partitioned phase space of a corresponding PADE underlying a QTG is shown in fine gray lines (dashed for threshold planes) underneath the QTG and allows identification of nodes representing singular resp. regular domains. In (a), STG of Ex. 1 and QTG of Ex. 2. In (b), the STG of a two component Boolean network with the corresponding QTG to the right. Heavier gray edges illustrate reachability properties discussed in the text.

First, we consider the number of attractors. In Fig. 1 (a) both systems have one attractor, but the STG in Fig. 1 (b) has two (a fixed point and a cyclic attractor) while the QTG has only one (a steady state in the upper right node). In Fig. 2 (a), the STG has fewer attractors than the corresponding QTG.

In addition, the relation between the attractor structure is not clear-cut. While the cyclic attractor of the STG in Fig. 1 (a) comprises all nodes and contains nodes with multiple outgoing edges, the cyclic attractor in the QTG is a simple cycle consisting only of two nodes joined by the heavier gray double edge in the lower part of the graph. In Fig. 1 (b) the cyclic attractor in the STG vanishes in the corresponding QTG. The same happens in Fig. 2 (b), but here an additional steady state can be observed in a singular node.

These examples illustrate that, in general, neither the number nor the character of the attractors is preserved. It can be shown that hyper-rectangular trap sets, i.e., node sets that no path can leave, correspond in the two graphs. This may be helpful in further elucidating the correspondences of attractors.

6   Discussion and Perspectives

In summary, the information inherent in the STG of the discrete update function is sufficient to derive the QTG of the corresponding PADE system and vice versa. Despite this fact, many characteristics of the two graphs are not preserved. This implies that, contrary to what might be expected, the QTG of the PADE system is not a straightforward refinement of the STG of the Thomas model.

Motivated by these findings, there are several directions for future work. First, we would like to better understand and characterize the network properties that lead to substantial differences, e.g., in the number of attractors in the dynamics of the Thomas and the PADE model. Second, we plan to extend the analysis

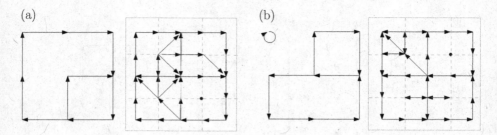

Fig. 2. Two examples for networks with two components and three activity levels for each component. In each case, the STG is depicted on the left, the corresponding QTG on the right. Depiction of the graphs corresponds to that in Fig. 1, only the explicit labeling of the STG nodes is omitted. In (a), the STG consists of a single cyclic attractor, while the QTG has an additional steady state at the lower right singular node depicted by a fat dot. In (b), both STG and QTG have a steady state in the upper left node. The STG has an additional cyclic attractor. The QTG has no cyclic attractor, but a singular steady state at the upper right singular node depicted by a fat dot.

to closely related formalisms like the refined qualitative representation of PADE systems [11] as well as piecewise multi-affine models [12]. Also, there exist approaches that allow the integration of threshold values directly into the Thomas formalism [13,14]. Clarifying the relation between the different approaches may allow one to transfer available results and analysis methods from one formalism to the other. Also, progress in this direction may be helpful when deciding on the most suitable and efficient modeling framework in a concrete application.

References

1. Thomas, R., D'Ari, R.: Biological Feedback. CRC Press (1990)
2. Snoussi, E.H.: Dyn. Syst. 4(3), 565–583 (1989)
3. Glass, L., Kauffman, S.L.: J. Theoret. Biol. 34(2), 219–237 (1972)
4. Glass, L., Kauffman, S.A.: J. Theoret. Biol. 39(1), 103–129 (1973)
5. de Jong, H., Gouzé, J.-L., Hernandez, C., Page, M., Sari, T., Geiselmann, J.: Bull. Math. Biol. 66(2), 301–340 (2004)
6. Snoussi, E.H., Thomas, R.: Bull. Math. Biol. 55(5), 973–991 (1993)
7. Chaves, M., Tournier, L., Gouzé, J.-L.: Acta Biotheoretica 58(2), 217–232 (2010)
8. Wittmann, D.M., Krumsiek, J., Saez-Rodriguez, J., Lauffenburger, D.A., Klamt, S., Theis, F.J.: BMC Systems Biology 3, 98 (2009)
9. Filipov, A.: Differential equations with discontinuous righthand sides. Springer, Heidelberg (1988)
10. Gouzé, J.-L., Sari, T.: Dyn. Syst. 17(4), 299–316 (2002)
11. Batt, G., de Jong, H., Page, M., Geiselmann, J.: Automatica 44(4), 982–989 (2008)
12. Kloetzer, M., Belta, C.: Transactions of the Institute of Measurement and Control 32(5), 445–467 (2009)
13. Thomas, R., Thieffry, D., Kaufman, M.: Bull. Math. Biol. 2(57), 24–276 (1995)
14. Richard, A., Bernot, G., Comet, J.-P.: Fund. Inform. 65, 373–392 (2005)

Automatic Inference of Regulatory and Dynamical Properties from Incomplete Gene Interaction and Expression Data

Fabien Corblin[1], Eric Fanchon[1], Laurent Trilling[1],
Claudine Chaouiya[2], and Denis Thieffry[3]

[1] UJF-Grenoble 1, CNRS, Laboratoire TIMC-IMAG, UMR 5525,
Grenoble, F-38041, France
{Fabien.Corblin,Eric.Fanchon,Laurent.Trilling}@imag.fr
[2] IGC - Instituto Gulbenkian de Ciência, Oeiras, Portugal
chaouiya@igc.gulbenkian.pt
[3] IBENS - UMR CNRS 8197 - INSERM 1024 - ENS, Paris, France
thieffry@ens.fr

Advanced mathematical methods and computational tools are required to properly understand the behavior of large and complex regulatory networks that control cellular processes. Since available data are predominantly qualitative or semi-quantitative, discrete (logical) modeling approaches are increasingly used to model these networks. Here, relying on the multilevel logical formalism developed by R. Thomas *et al.* [7,9,8], we propose a computational approach enabling (i) to check the existence of at least one consistent model, given partial data on the regulatory structure and dynamical properties, and (ii) to infer properties common to all consistent models. Such properties represent non trivial deductions and could be used by the biologist to design new experiments. Rather than focusing on a single plausible solution, *i.e.* a model fully defined, we consider the whole class of models consistent with the available data and some economy criteria, from which we deduce shared properties. We use constraint programming to represent this class of models as the set of all solutions of a set of constraints [3]. For the sake of efficiency, we have developed a framework, called SysBiOX, enabling (i) the integration of partial gene interaction and expression data into constraints and (ii) the resolution of these constraints in order to infer properties about the structure or the behaviors of the gene network. SysBiOX is implemented in ASP (Answer Set Programming) using Clingo [4].

We apply this approach to the regulatory network controlling the earliest steps of *Drosophila* embryo segmentation, *i.e.* the gap genes and their cross-regulations, under the additional control of maternal gene products [6,5,1,2,3]. We consider three kinds of data.

First, published molecular genetic studies enable the identification of the main actors as well as the establishment or the suggestion of cross-regulatory interactions. The actors we consider are the transcription factors Hb, Kr, Kni, Gt, and the maternal factors Cad and Bcd, to which we add a terminal control factor, Ter, which represents the effect of a combination of several terminal factors. The network of interactions is represented in Figure 1.

M.A. Lones et al. (Eds.): IPCAT 2012, LNCS 7223, pp. 25–30, 2012.
© Springer-Verlag Berlin Heidelberg 2012

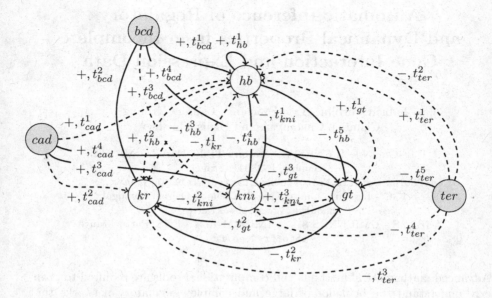

Fig. 1. Generic structure of the regulatory network controlling the first steps of drosophila embryo segmentation, where solid arcs denote well-established interactions, whereas dotted arcs denote potential interactions. Each arc is associated with a symbolic threshold and a sign (to account for positive versus negative regulations). Grey nodes denote input components (maternal factors).

Next, we consider qualitative information on the spatio-temporal expression profiles of the main genes involved in the process. In this respect, we can rely on the information already integrated in the database FlyEx[1]. For example, Figure 2-A displays the expression profiles of the main gap and maternal gene products along the antero-posterior (AP) axis of the embryo around the 13th nuclear division cycle. At that time, the gap genes are expressed in one or two peaks organized in a well-defined sequence along the AP axis. We integrate these spatial characteristics using a qualitative description in terms of constraints on successions of extrema for the gap genes. As a first approximation, we consider that the expression state reached by each cell at the end of the 13th division cycle corresponds to a stable state, which varies along the AP axis. More precisely, based on the profiles observed in the wild type situation, we enforce a pattern of successive stable states along the anterior-posterior axis of the considered embryonic region (which corresponds to the trunk and caudal regions of the embryo, the head region being subject to more complex and partly unknown regulations): a decrease of Gt, followed by a decrease of Hb, a peak of Kr (an increase followed by a decrease), a peak of Kni, a peak of Gt, a peak of Hb, and finally an increase of Ter. To account for the wild-type pattern shown in Figure 2-B, first row, we consider 7 regions (each corresponding to a stable expression

[1] http://urchin.spbcas.ru/flyex/

state). We further rely on the data available on the gap gene expression profiles for seven loss-of-function mutations, affecting maternal or gap genes. In this respect, we consider 5 regions for the type *hb*0 (loss-of-function of Hb), 3 for *kr*0, 5 for *kni*0, 4 for *gt*0, 3 for *bcd*0, 5 for *cad*0 and 5 for *ter*0. For each mutant, the qualitative features of the profiles are described in a separate row in Figure 2-B.

Finally, for the wild type, we also introduce constraints specifying that the stable pattern is reachable from relevant initial states. The Figure 2-C represents the constraints on these initial states : no expression of Kr, Kni and Gt; positive gradients for Cad and Ter, and negative gradients for Hb and Bcd. We enforce a maximal length of 6 states for this reachability path in order to avoid back and forth updating of the same factors. In summary, we enforce (i) the existence of 7 reachability paths, one for each segment of the wild type, each with a length smaller or equal to 6, and (ii) the equality, in each region, of the last state of the path with the corresponding stable state.

On the basis of this combination of interaction and expression constraints, the challenge is to identify the minimal complying model(s), *i.e.* the model(s) involving all established regulatory edges, along with a minimal set of potential ones, while minimizing the number of distinct thresholds. In addition, we aim at identifying properties of discrete kinetic parameters. These parameters come from the logical formalism of R. Thomas [7,9,8] and specify the target discrete value of a component when subject to a combination of interactions.

The consistency of the data (*i.e.* the existence of at least one consistent model) is proved with SysBiOX in 3338 seconds, using a Linux PC with an Intel Core2Duo processor at 2.4GHz with 2.9GB of RAM.

We choose to follow the following 3-steps strategy to query the set of partial data (proved consistent):

1. Minimize the number of potential interactions;
2. Minimize the number of threshold values;
3. Infer properties on parameters.

Successively applying these queries, we first obtain (in 1016 seconds) a unique minimal regulatory graph, which includes only two potential interactions (inhibition of Kr by Hb, and inhibition of Kni by Ter, see Figure 3-A).

Next, we obtain (in 368 seconds) a unique threshold instantiation, which minimizes the number of theshold values per component (*i.e.* which minimizes the number of distinct discrete values to be associated to each regulatory component). Consequently, we obtain a unique regulatory graph that minimizes both the number of interactions and the number of distinct thresholds. This unicity was not expected at all.

Finally, we find properties relating the logical parameters (fixed values or inequalities) of the model, which are summarized in Figure 3-B.

In conclusion, this study clearly demonstrates that a constraint based approach makes it possible (i) to enforce structural and behavioral partial information about a regulatory network, and (ii) to define a variety of functionalities to address questions about network structure, thresholds, kinetic parameters,

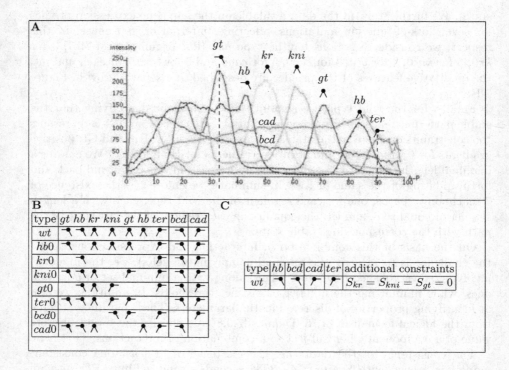

Fig. 2. (A) Expression profiles of the main gap and maternal gene products along the anterior-posterior axis around the 13th nuclear division cycle. (B) Qualitative spatial description of these profiles, for the wild-type (1st row) and published mutants, e.g. *hb0* denoting hb loss-of-function. Note that genes Bcd and Cad are not subject to spatial requirements depending on oter genes. (C) Qualitative spatial description of the initial states S for the wild-type.

behaviors. In addition to this rich expression power, the constraint approach presents the computational advantage (in order to automatically infer properties) of considering the entire set of alternative models in "intention" (solutions of a set of constraints) rather then by extension (by enumerating the models). The set of solutions inferred for the gap regulatory module still needs to be analyzed in greater details. Selected model solutions will be simulated for varying genetic backgrounds and compared to experimental data. Depending on simulation results, additional constraints may be considered to improve the predictability of our approach in order to design novel informative experiments.

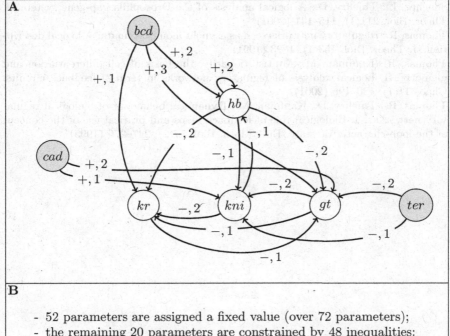

A

B

- 52 parameters are assigned a fixed value (over 72 parameters);
- the remaining 20 parameters are constrained by 48 inequalities:
  - 12 inequalities relate one parameter and one threshold,
  - 36 inequalities relate two parameters.

Fig. 3. Summary of the results: (A) The minimal interaction graph shared by all minimal inferred models, with fixed interactions and threshold values. (B) Some properties of the logical parameters shared by all the resulting consistent models.

References

1. Alves, F., Dilão, R.: Modeling segmental patterning in drosophila: Maternal and gap genes. J. Theor. Biol. 241(2), 342–359 (2006)
2. Ashyraliyev, M., Siggens, K., Janssens, H., Blom, J., Akam, M., Jaeger, J.: Gene circuit analysis of the terminal gap gene huckebein. PLoS Comput. Biol. 5(10), e1000548 (2009)
3. Corblin, F., Fanchon, E., Trilling, L.: Applications of a formal approach to decipher discrete genetic networks. BMC Bioinformatics 11, 385 (2010)
4. Gebser, M., Kaminski, R., Kaufmann, B., Ostrowski, M., Schaub, T., Schneider, M.: Potassco: The Potsdam answer set solving collection. AI Communications 24(2), 105–124 (2011)
5. Jaeger, J., Blagov, M., Kosman, D., Kozlov, K.N., Manu, M.E., Surkova, S., Vanario-Alonso, C.E., Samsonova, M., Sharp, D.H., Reinitz, J.: Dynamical analysis of regulatory interactions in the gap gene system of drosophila melanogaster. Genetics 167(4), 1721–1737 (2004)

6. Sánchez, L., Thieffry, D.: A logical analysis of the Drosophila gap-gene system. J. Theor. Biol. 211(1), 115–141 (2001)
7. Thomas, R.: Regulatory networks seen as asynchronous automata: A logical description. J. Theor. Biol. 153(1), 1–23 (1991)
8. Thomas, R., Kaufman, M.: Multistationarity, the basis of cell differentiation and memory –II. Logical analysis of regulatory networks in term of feedback circuits. Chaos 11(1), 180–195 (2001)
9. Thomas, R., Thieffry, D., Kaufman, M.: Dynamical behaviour of biological regulatory networks – I. Biological role of feedback loops and pratical use of the concept of the loop-characteritic state. Bull. Math. Biol. 57(2), 247–276 (1995)

CRISPR Transcript Processing:
An Unusual Mechanism for Rapid Production
of Desired Molecules

Marko Djordjevic[1], Konstantin Severinov[2], and Magdalena Djordjevic[3]

[1] Institute of Physiology and Biochemistry, Faculty of Biology,
University of Belgrade, Serbia
`dmarko@bio.bg.ac.rs`
[2] Waksman Institute for Microbiology, Rutgers University, USA
[3] Institute of Physics Belgrade, University of Belgrade, Serbia

Abstract. CRISPR is a recently discovered adaptive prokaryotic immune system. A crucial step in CRISPR defense mechanism is transcription of CRISPR cassette, which is followed by processing of the resulting long transcript (pre-crRNA) into small RNA molecules (crRNA) that recognize invading viruses. We model CRISPR transcript processing, and show that the system functions as a strong amplifier, which can rapidly generate a large number of crRNAs from only few pre-crRNA molecules. Based on this analysis, we propose a synthetic gene circuit that can produce a large number of desired molecules from a potentially toxic substrate.

1 Background

CRISPR (Clustered Regularly Interspaced Short Palindromic Repeats) cassettes consist of identical direct repeats of about 30 bp in length, interspaced with variable spacers of similar length [1]. It was recently discovered that CRISPR presents an adaptive prokaryotic immune system, which is responsible for defending prokaryotic cell against invaders, so that a match between a CRISPR spacer and invading phage (bacterial virus) sequence provides immunity to infection [2]. In addition to CRISPR cassettes, CRISPR-associated (cas) genes are also required for this immunity. Experiments show that the entire CRISPR locus is transcribed as a long transcript (called pre-crRNA), which is further processed by one of the Cas proteins to yield small RNAs (called crRNAs); crRNAs are responsible for recognition and - together with Cas proteins - inactivation of invading viruses [3].

CRIPSR/Cas system in E. coli is silent under normal growth conditions, and mechanism of the system induction upon virus infection is currently unclear [1]. Surprisingly, when cas genes are (artificially) overexpressed in cell, several thousand crRNA molecules are generated from only few pre-crRNA molecules [4]. Furthermore, while overexpression of only cas genes or CRISPR cassette leads to only partial protection against invading viruses [5], their joint overexpression leads to the complete immunity [3].

M.A. Lones et al. (Eds.): IPCAT 2012, LNCS 7223, pp. 31–34, 2012.

2 Results

To achieve a quantitative understanding of the experimental results, we developed a model of CRISPR processing, which we analyzed both deterministically and stochastically [6]. We show that, upon cas overexpression, the system acts as a strong amplifier, so that a small decrease of pre-crRNAs results in a large increase of crRNAs; such strong amplification is consistent with the experimental observations. Surprisingly, this strong amplification crucially depends on fast nonspecific degradation of pre-crRNA by an yet unspecified endonuclease. We furthermore find that overexpression of cas above certain level does not result in an additional increase of steady-state crRNA levels. However, a joint overexpression of CRISPR and cas genes overcomes this saturation in steady-state crRNA levels, which provides a plausible explanation for why such joint overexpression is necessary for complete immunity [3,5]. Moreover, this joint overexpression leads to a 'burst' of crRNA, so that hundreds of crRNAs are generated in a small time interval upon system induction (Figure 1).

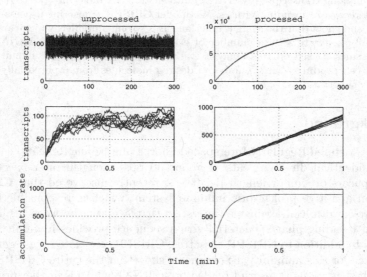

Fig. 1. Burst of crRNAs upon joint overexpression of CRISPR and cas genes. The red curve and the blue curves correspond, respectively, to the deterministic, and to ten stochastic trajectories. The left and the right panels correspond, respectively, to pre-crRNAs and crRNAs. The first and the second row correspond, respectively, to transcript increase, during first 300min, and during first 1min, upon system induction. Notice that crRNA has a much larger increase compared to pre-crRNA. The third row corresponds to the rate of transcript accumulation during the first minute upon system induction. Notice the two order of magnitude increase in the transcript accumulation rate in a short time interval (the burst of crRNA) [6]. Parameters in the figure correspond to (artificial) overexpression of CRISPR cassette and cas genes, as in [3,5].

Inspired by our study of CRISPR/Cas system in E. coli, we next investigate design of a synthetic gene circuit that can generate a large amount of useful product from small amounts of potentially toxic substrate [7]. Analysis of such circuit is motivated by the ability of CRISPR/Cas system in E. coli to generate a large number of crRNAs from only few pre-crRNA molecules. We optimize the circuit, with the goal of generating maximal product amounts, without increase of substrate amounts upon system induction. It is interesting that the derived optimal values of substrate and product decay rates roughly match experimentally inferred pre-crRNAs and crRNAs decay rates. We furthermore show that the optimal system promotes a fast transition to a large number of product molecules, which is due to the rapid degradation of the substrate molecules. While the parameters/conditions of natural CRISPR/Cas system induction are currently unclear, it will be interesting to compare how closely they correspond to those of the optimal system, once experimental data on the natural system induction become available.

3 Conclusion

We here developed the first quantitative model of CRISPR transcript processing, which is able to explain all existing steady state experimental measurements. Furthermore, the model indicates a surprisingly large burst of crRNA immediately upon the system induction; this result may establish joint activation of CRISPR and cas genes as the mechanism to exploit a short window of opportunity for neutralizing an incoming virus. Inspired by the mechanism of CRIPR/Cas transcript processing, we proposed a novel synthetic gene circuit that can rapidly produce a large amount of useful molecules from substrate that has to be kept in small amount.

Acknowledgements. This work is partially supported by Marie Curie International Reintegration Grant within the 7th European community Framework Programme (PIRG08-GA-2010-276996) and by the Ministry of Education and Science of the Republic of Serbia under project number ON173052.

References

1. Al-Attar, S., Westra, E.R., Van der Oost, J., Brouns, S.J.J.: Clustered regularly interspaced short palyndromic repeats (CRISPRs): the hallmark of an ingenious antiviral defense mechanism in prokaryotes. Biol. Chem. 392, 277 (2011)
2. Barrangau, R., et al.: CRISPR provides acquired resistance against viruses in prokaryotes. Science 315, 1709 (2007)
3. Brouns, S.J.J., et al.: Small CRISPR RNAs guide antiviral defense in prokaryotes. Science 321, 960 (2008)
4. Pougach, K., Semenova, E., Bogdanova, E., Datsenko, K.A., Djordjevic, M., Wanner, B.L., Severinov, K.: Transcription, processing and function of CRISPR cassettes in Escherichia coli. Mol. Microbiol. 77, 1367 (2010)

5. Westra, E.R., et al.: H-NS-mediated repression of CRISPR based immunity in Escherichia coli K12 can be relieved by the transcription activator LeuO. Mol. Microbiol. 77, 1380 (2010)
6. Djordjevic, M., Severinov, K.: CRISPR transcript processing: an unusual mechanism for rapid generation of small RNAs. Submitted to J. Theor. Biol.
7. Djordjevic, M., Djordjevic, M.R.: A synthetic gene circuit for rapid product generation from small substrate amounts. Submitted to Phys. Rev. Lett.

A Comprehensive Computational Model to Simulate Transcription Factor Binding in Prokaryotes

Nicolae Radu Zabet[1,2] and Boris Adryan[1,2]

[1] Cambridge Systems Biology Centre, University of Cambridge,
Tennis Court Road, Cambridge CB2 1QR, UK
[2] Department of Genetics, University of Cambridge, Downing Street,
Cambridge CB2 3EH, UK
n.r.zabet@gen.cam.ac.uk,
adryan@sysbiol.cam.ac.uk

Site specific transcription factors (TF) are proteins that orchestrate transcription by binding to specific target sites on the DNA. This binding can be both sequence- and conformation-specific. However, also non-specific binding with lower affinity can be observed [3]. The number of specific target sites is significantly smaller compared to the number of non-specific sites and, consequently, TF molecules bind, in a first instance, non-specifically to the DNA. Once bound to the DNA the TF molecules perform an one dimensional random walk on the DNA until they either find a target site or unbind from the DNA template. In particular, during the one dimensional random walk on the DNA, a molecule will perform one of the three types of movements: (*i*) sliding , (*ii*) hopping and (*iii*) jumping [6]. This combination of one and three dimensional diffusion is called *facilitated diffusion* and it is hypothesised that this speeds up the search process [3,2,5].

Most theoretical works focussed on analytical solutions of the facilitated diffusion mechanism [6]. These analytical solutions provided a detailed overview of the process and new insights into the underlining mechanisms of the search process. Nevertheless, analytical solutions are not able to incorporate real DNA sequences, as they do not have homogeneous affinity landscapes, and cannot model molecular crowding with dynamic obstacles.

Here we will present a comprehensive computational framework that allows the stochastic simulation of the search process of TFs for their target sites on the DNA. Each TF molecule is represented as an object (agent), which can move through three dimensional diffusion in the bacterial cytoplasm, but which also can bind to the DNA and perform an one dimensional random walk.

One problem with this type of computational models is the trade-off between the level of detail of the system and the simulation speed, i.e., a detailed representation of the system can lead to slower performance while abstract representation can miss essential biological aspects of the system. With our solution, the analysed system can be extremely large. For example, *E.coli* K-12 has a 4.6 *Mbp* genome and there are $\sim 10^4$ DNA binding proteins. In this contribution, we will

M.A. Lones et al. (Eds.): IPCAT 2012, LNCS 7223, pp. 35–37, 2012.

present a novel and highly efficient implementation of the TF search process, which simulates significantly faster that previous models [4,1].

First, we want to demonstrate how the molecules move on the DNA during a simulation run. Fig. 1 shows an example of a random walk performed by 1 or 3 molecules on a 250 *bp* randomly generated DNA sequence. The molecules alternate the one dimensional movements with three dimensional excursions or hops.

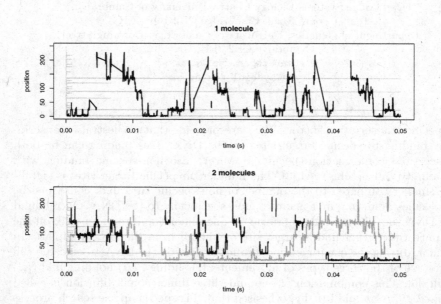

Fig. 1. *Dynamic Behaviour of TF molecules.* We consider a random 250 *bp* DNA sequence and TF molecules which can bind/unbind, hop, jump, slide left/right. (**Top**) 1 TF molecule (**Bottom**) 2 TF molecules. The position of the molecules is represented on y-axis and the time on the x-axis. The grey line on the y-axis represents the affinity at that position for a TF.

Furthermore, we also investigate the effects of facilitated diffusion on the occupancy-bias. In the case of only one molecule, one can observe strong positive correlation between site occupancy on the DNA and affinity for a particular sequence. Furthermore, in a crowded environment, some low affinity sites display high occupancy (false positives), while some high affinity sites display low affinity (false negatives), see Fig. 2.

Our model represents an ideal entry point for stochastic simulations on transcription factor target finding in prokaryotic systems. In addition, we provide an implementation in Java 1.6 of this computational model, which is available at http://logic.sysbiol.cam.ac.uk/grip/. The implementation will allow researchers not only to use the model with user-defined parameter sets, but also

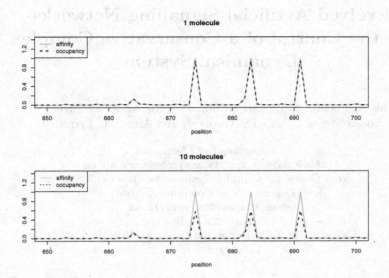

Fig. 2. *Affinity vs Occupancy.* We consider a random 1000 *bp* DNA strand. In the top graph we show the normalized affinity and normalized occupancy for 1 molecule and in the bottom graph, for 10 molecules.

to extent and adjust the model to their needs. Finally, we are going to present further results on facilitated diffusion obtained with our computational model at the conference.

References

1. Barnes, D.J., Chu, D.F.: An efficient model for investigating specific site binding of transcription factors. In: 2010 4th International Conference on Bioinformatics and Biomedical Engineering (iCBBE), June 18-20, pp. 1–4. IEEE Xplore, Chengdu (2010)
2. Benichou, O., Loverdo, C., Moreau, M., Voituriez, R.: Optimizing intermittent reaction paths. Physical Chemistry Chemical Physics 10(47), 7059–7072 (2008)
3. Berg, O.G., Winter, R.B., von Hippel, P.H.: Diffusion-driven mechanisms of protein translocation on nucleic acids. 1. models and theory. Biochemistry 20(24), 6929–6948 (1981)
4. Chu, D., Zabet, N.R., Mitavskiy, B.: Models of transcription factor binding: Sensitivity of activation functions to model assumptions. Journal of Theoretical Biology 257(3), 419–429 (2009)
5. Halford, S.E.: An end to 40 years of mistakes in dnaprotein association kinetics? Biochemical Society Transactions 37, 343–348 (2009)
6. Mirny, L., Slutsky, M., Wunderlich, Z., Tafvizi, A., Leith, J., Kosmrlj, A.: How a protein searches for its site on dna: the mechanism of facilitated diffusion. Journal of Physics A: Mathematical and Theoretical 42, 434013 (2009)

Evolved Artificial Signalling Networks for the Control of a Conservative Complex Dynamical System

Luis A. Fuente[1,2], Michael A. Lones[1,2], Alexander P. Turner[1,2],
Susan Stepney[1,3], Leo S. Caves[1,4], and Andy M. Tyrrell[1,2]

[1] Department of Electronics
{laf509,mal503,apt503,amt}@ohm.york.ac.uk
[2] York Centre for Complex Systems Analysis (YCCSA)
[3] Department of Computer Science
susan.stepney@cs.york.ac.uk
[4] Department of Biology
University of York, Heslington, York, YO10 5DD, UK
lsdc1@york.ac.uk

Abstract. Artificial Signalling Networks (ASNs) are computational models inspired by cellular signalling processes that interpret environmental information. This paper introduces an ASN-based approach to controlling chaotic dynamics in discrete dynamical systems, which are representative of complex behaviours which occur in the real world. Considering the main biological interpretations of signalling pathways, two ASN models are developed. They highlight how pathways' complex behavioural dynamics can be captured and represented within evolutionary algorithms. In addition, the regulatory capacity of the major regulatory functions within living organisms is also explored. The results highlight the importance of the representation to model signalling pathway behaviours and reveal that the inclusion of crosstalk positively affects the performance of the model.

1 Introduction

Cells need to engage in many forms of communication in order to sense and respond to the outside world. This capacity is vital for cells to survive. Cellular signalling involves a chain of events that permits cells to interact with their environment. It starts with the triggering of a biochemical signal and terminates with an adaptive cellular response. Classically, cellular signalling may be conceived as follows: a surface receptor binds an extracellular messenger (e.g. hormone, growth factor), which diffuses an intracellular signal to an effector protein inside the cell. This then produces secondary messengers, which transmit the information further into the cell. Spatially or temporally variable catalytic reactions or cascades of protein kinases finally lead to changes in gene expression, bringing about a change in cellular activity.

M.A. Lones et al. (Eds.): IPCAT 2012, LNCS 7223, pp. 38–49, 2012.

In this paper we propose a new Artificial Signalling Network (ASN) approach to modelling the spatial properties and temporal topologies of cellular signalling, capturing its intrinsic dynamics. In order to test the model we apply it to the control of a numerical dynamical system, whose properties mirror the complexity of cellular environments. Controlling dynamics also represents a classical multi-disciplinary problem in its own right.

This paper is organised as follows: Section 2 presents a brief overview of dynamical systems, Section 3 reviews the current literature on ASNs, Section 4 introduces the new model and defines its methodology, Section 5 presents some initial results, and Section 6 concludes.

2   Dynamical Systems

A dynamical system is a mathematical model where a function, or *evolution rule*, characterises its state based on the system's current state and initial conditions [12]. The evolution rule defines the motion and behaviour of the system across the state space. Dynamical systems are initially divided into *autonomous* and *non-autonomous*. The former is a closed system whose dynamics are not perturbed by the outside world. The latter defines an open system changing over time, as inputs are received from an external environment. Likewise, dynamical systems can be *discrete* or *continuous* in time, depending on the type of evolution rule: difference equations in the former and differential equations in the latter.

Given a set of initial points within a discrete state space, the evolution rule defines their *trajectories* as a sequence of states over a period of time. A dynamical system where trajectories do not contract to a limited region of the state space is known as a *conservative* system.

Dynamical systems can display a wide range of behaviours. The most interesting are those involving holistic irregular and unpredictable properties; this atypical dynamism is known as *chaos*. Despite being deterministic, chaotic systems display aperiodic behaviours characterised by an exponential sensitivity to initial conditions and the existence of strange attractors. Whereas the former suggests that small changes in the initial conditions convey highly different trajectories throughout the state space, the latter defines fractal and non-linear regions where trajectories may converge.

2.1   Chirikov's Standard Map

Chirikov's standard map [4] is a conservative and discrete two-dimensional dynamical system representing iteratively the interactions of two canonical variables within the unit square:

$$x_{n+1} = (x_n + y_{n+1}) \bmod 1 \qquad y_{n+1} = y_n - \frac{k}{2\pi} \sin(2\pi x_n) \qquad (1)$$

One of the map's main properties is its capacity to represent different dynamics as its nonlinearily increases. Thus, low values of k preserve an ordered state where trajectories lead to periodic and quasi-periodic trajectories bounded on the y-axis (see Fig.1(a)). As k increases, chaotic dynamics arise in the form of chaotic islands along the y-axis, which are never visited (see Fig.1(b)). The type of trajectories depends on the map's initial conditions. The map shows a behavioural inflection point, k_c, at $k \approx 0.972$. Initial impermeability progressively disappears as $k > k_c$ (see Fig. 1(c)–(d)), enabling trajectories to vertically travel across the map. The example in Figure 1 shows the permeability of the map increasing as k increases, characterised by the gradual encroachment of the chaotic regions.

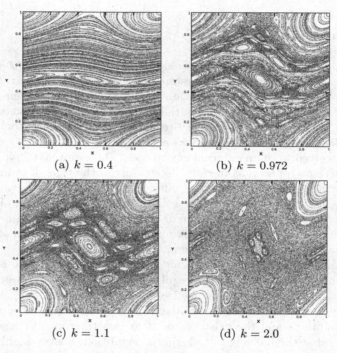

(a) $k = 0.4$ (b) $k = 0.972$

(c) $k = 1.1$ (d) $k = 2.0$

Fig. 1. Sampled trajectories of Chirikov's standard map using different values of k, showing the change from the ordered to the chaotic state. Each map is plotted using 400 randomly chosen initial points across the unit interval over 800 iterations.

2.2   State Space Targeting

The sensitivity underlying chaotic dynamics brings about a broad band of complex, unstable and unpredictable behaviours using arbitrary control signals. *Controlling chaos* or *chaos targeting* attempts to provoke large behavioural changes in the dynamics using relatively small perturbations, which are achievable by the modification of the system's control signals. Therefore, it looks at finding the fastest path from an initial condition to a target point. Existing research has shown that targeting in conservative systems, such as Chirikov's standard map, is achievable using small perturbations to drive across the different chaotic regions of the state space [9,14]. Consequently, the map becomes navigable from the bottom to the top and it is possible to find a controller able to transverse it.

3   Artificial Signalling Networks

As an abstraction of cellular signalling, Artificial Signalling Networks try to model the particular characteristics that allow cells to take chemical signals as inputs and generate some adaptive output. Given our motivation to preserve biological plausibility, we are interested in investigating the ASN's ability to implement control functionalities. From a computational perspective, the importance of ASNs lies in their capacity to coordinate the set of events within cells that trigger robust, efficient and specific responses, their ability to work as independent processing units and their capacity to adapt to environmental perturbations.

One way to model ASNs relies on the quantitative description of particular pathways. Experimental and mathematical approaches facilitate the identification of the functional elements as well as their interactions in concrete pathways, thereby simplifying their modelling. Thus, the fuzzy model suggested in [7] computes the dynamics of the IL-6 pathway based on the state of the components, the initial inputs and a set of fuzzy rules. Likewise, the validity of logic-based modelling has been widely demonstrated [10]. These models have a direct physical basis. Said et al. in [13] take a more abstract approach, modelling the interaction between two participating elements, and then, applying it to simulate MAP kinase cascade as a Markov chain. However, the reconstruction of these pathways was insufficiently accurate since the complexity displayed by some of the components could not be captured.

Another way to design ASNs is to use evolutionary algorithms. They can induce complex behaviours in a concise and evolvable way [9] and some specific functionalities are achievable only through evolutionary processes [6]. In fact, evolved ASNs have been successfully used to capture simple forms of biological signal processing [3,5]. In this paper we propose an alternative approach: we use a generic evolved artificial signalling network, where no specific information of either the participating elements or their interactions is needed. Therefore, all limitations emerging from the pathway's particular characteristics are dismissed and the ASN's topology is the result of its interaction with the environment. This increases its adaptability when facing different types of environment. A similar approach has been suggested in [8]. However, it models ASNs as Boolean networks and limits the connectivity between the participating elements.

4   State Space Targeting with ASNs

Most of the signalling processes inside the cells involve complex interactions between enzymes. Although these interactions may vary in size, they are essential in the transmission of signals. In practice, enzymes are not functional unless they are grouped together into a biological structure. Likewise, some of the main cellular functions are only achievable under certain spatial distributions. There are a wide variety of abstractions aiming to represent the properties of intracellular signalling networks. However, many of them fail to fulfill this objective, or it is only partially achieved. For example, Bayesian Networks [11,15], limit the representation of complex dynamics due to their acyclic nature. This paper proposes the usage of *interaction graphs* to capture the topological and temporal patterns intrinsic to signalling pathways. An interaction graph is a mathematical representation modelling the dynamical behaviours of a system formed by multiple actors interacting over time; thus, we consider ASNs as dynamical systems defined by interactions between enzymes.

According to the different types of pathways inside cells, we introduce two approaches for ASN modelling. Whilst the first model considers ASNs as a subtype of metabolic network (see Fig. 2(a)), the second considers them to be cascades of protein kinases (see Fig. 2(b)). Both approaches extend the model described in [9]. Both are continuous-valued models as this enables a more realistic representation of biological systems. To allow a valid comparison between both models, they are deterministic and synchronous.

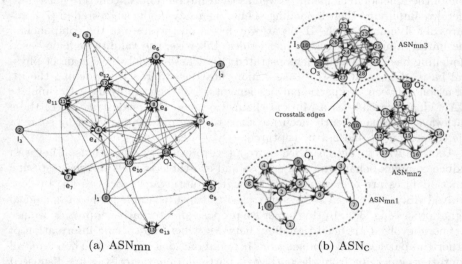

(a) ASN_{mn} (b) ASN_C

Fig. 2. Representation of both ASN models. Each models has three inputs, I_i, and one global output O_i. ASN_{mn} contains 15 enzymes. ASN_C has 3 ASN_{mn}, each of which has 10 enzymes. Crosstalk edges are the arcs connecting two ASN_{mn}s.

The artificial signalling network as a metabolic network. (ASN_{mn}) is defined as a directed interaction graph, where the nodes are an indexed set of enzymes and the edges represent their biochemical reactions. Every enzyme contains a set of substrates, a set of products and a mapping function relating the concentrations of both sets of chemicals. Formally: $ASN_{mn} =< C, E, R, I_E, O_E >$, where:

C is the indexed set of random chemical concentrations $\{c_0, c_1, \ldots, c_n : \mathbb{R}\}$.

E is the indexed set of enzymes $\{e_0, e_1, \ldots, e_n \ : \ e_i =<S_i, P_i, m_i>\}$, where:

$S_i \subseteq C$ is the concentration of the substrates used by the enzyme.

$P_i \subseteq C$ is the concentration of the products generated by the enzyme.

$m_i : \mathbb{R}^n \to \mathbb{R}^n$ is the enzymes' substrate-product mapping function.

R defines the set of enzymatic reactions $\{r_0, r_1, \ldots, r_n \ : \ r_i \in \{+, -\}\}$. Negative and positive values indicate enhancing and inhibition respectively.

$I_E \subset E$ is the set of enzymes used as inputs.

$O_E \subset E$ is the set of enzymes used as outputs.

The execution of the ASN proceeds as follows:

1. S_i and P_i are randomly initialised (if ASN not previously executed).
2. The concentrations of S_i in I_E are set by the external inputs.
3. At each time step, each enzyme e_i applies its mapping function m_i to determine the new concentration of the products P_i based on the concentration of its substrates S_i. In the particular case where the substrate is provided by multiple enzymes, the new concentration is the mean output of all different contributing enzymes.
4. After a number of time steps the execution is halted and the concentrations of the products in O_E are copied to the external outputs.

The artificial signalling network as a set of protein kinases cascade. ($\mathrm{ASN_C}$) extends the previous model by grouping the enzymes into an indexed set of $\mathrm{ASN_{mn}}$, each of which represents a signalling pathway. Additionally, pathway crosstalk is simulated by a set of edges connecting two $\mathrm{ASN_{mn}}$s. Formally $\mathrm{ASN_C} =< ASN, C_R, O_E, I_E >$, where:

ASN is an indexed set of artificial signalling networks $\{asn_0, asn_1, \ldots, asn_n \ : \ asn_i \equiv ASN_{mn} =< C, E, R, I_E, O_E >\}$.

$I_E \subset E$ is the set of enzymes used as inputs, where $|ASN_{mn}| = |I_E|$.

$O_E \subset E$ is the set of enzymes used as outputs.

$C_R \in [0, 1]$ is the probability of crosstalk.

The execution of $\mathrm{ASN_C}$ is similar to that of $\mathrm{ASN_{mn}}$:

1. S_i and P_i are randomly initialised (if ASN not previously executed).
2. The concentrations of S_i in I_E are set by the external inputs. Each $\mathrm{ASN_{mn}}$ has only one input.
3. At each time step, each enzyme e_i applies its mapping function m_i to determine the new concentration of the products P_i based on the concentration of its substrates S_i. When the substrate is provided by multiple enzymes, the new concentration is the mean output of all different contributing enzymes. Additionally, enzymes having a crosstalk edge have their products asymptotically reduced to half of their maximum rate.
4. After a number of time steps ($t_s \in [1, 100]$) the execution is halted and the external outputs are calculated as the mean output value of all contributing networks.

We also analyse the effect of having two types of enzymes depending on the number of times they are phosphorylated (single and double). This more closely represents the different phosphorylation states in cascades of protein kinases.

4.1 Mappings

Three types of parameterisable functions are chosen as enzyme mappings: the Hill, the Michaelis-Menten and the first-order kinetics equations. They are the most common models of molecular regulatory functions within living organisms.

The Hill equation describes the cooperative level between an enzyme and its substrate as $f(x) = v|x|^h/k^h|x|^h$, where $v \in [0,1]$ is the asymptotic threshold, $k \in [0,1]$ determines its gradient and $h \in \mathbb{R}^n$ is the hill coefficient indicating the degree of cooperativeness. The equation can also be extended by adding the probability of binding $\beta \in [0,1]$. If $h = 1$ the Hill equation is equivalent to the Michaelis-Menten equation. For multiple inputs, $x = \sum_{j=0}^{n} \frac{i_j w_j}{n}$. Negative values indicate inhibition and $f(x)^- = 1 - f(x)$.

The Michaelis-Menten equation characterises the enzyme kinetic reactions. It is a hyperbolic function $f(x) = v|x|/(k - |x|)$, where $v \in [0,1]$ is the asymptotic threshold and $k \in [0,1]$ determines its gradient. For multiple inputs, $x = \sum_{j=0}^{n} \frac{i_j w_j}{n}$, where $i_0 \ldots i_n$ are inputs and $w_0 \ldots w_n \in [-1,1]$ are the corresponding input weights. Negative values indicate inhibition and $f(x)^- = 1 - f(x)$.

The Multi-Dimensional Michaelis-Menten equation defines the enzymes' kinetics when substrates are produced by multiple enzymes based on the probability of binding as $f(x) = \sum_{i=0}^{k} \beta_i (x_i/k_i)^{n_i} / 1 + \sum_{i=0}^{n} (x_i/k_i)^{m_i}$, where $\beta \in [0,1]$ is the binding probability, $v \in [0,1]$ is the asymptotic threshold and $k \in [0,1]$ determines its gradient. $m, n \in \mathbb{R}^n$, where $m = n$ for activation and $n = 0$ and $m > 0$ for repression [1].

The first-order kinetics equation is the simplest kinetics model relating to the rate of phosphorylation of an enzyme to the concentration of its active site and the concentration of the unphosphorylated substrate. When single phosphorylated $f(x) = v|x|/(1 + |x|)$, and when double $f(x) = vx^2/(1 + |x| + x^2)$, where $v \in [0,1]$ is the asymptotic threshold. For multiple inputs, $x = \sum_{j=0}^{n} \frac{i_j w_j}{n}$. Negative values indicate inhibition and $f(x)^- = 1 - f(x)$.

4.2 Methodology

Both ASN models have been evolved using a standard generational evolutionary algorithm with tournament selection (size=4), uniform crossover (rate = 0.48), and point mutation (rate = 0.16). 40% of the solutions are mutated. An ASN is encoded as an array of genes, followed by an integer within the interval

[0, 100] representing the number of time steps for execution. Crossover points lie between the enzymes' boundaries. In an attempt to reduce the complexity of the analysis, the number of genes and enzymes has been fixed at 10. All runs terminate after 100 generations.

Initial chemical values and mapping parameters are represented using floating-point values and mutated using a Gaussian function with its center at the current values. However, mutation is constrained to one of the following operations to fulfill the restrictions proposed in [16] to model the reactions between molecules:

1. Increasing or decreasing the chemical values.
2. Changing the state of the biochemical reactions by modifying the parameters of the mapping functions.
3. Variation of the reaction rates by changing their weight values.
4. Adding or removing participants (edges) to/from the reactions.
5. Variation of the number of time steps.

Each ASN is represented as an interaction graph in which each vertex is an enzyme. External inputs, which represent the controller's state space location at the start of execution, are always delivered to the inputs of low-numbered enzymes (in terms of the network's genetic encoding). External outputs, which determine the new values for control signals, are always read from the outputs of high-numbered enzymes.

Traversing Chirikov's standard map: The goal is to evolve an ASN-based controller which can guide trajectories from a designated region at the bottom of the map to a designated region at the top of the map by modulating the control signal k within the range $[1.0, 1.1]$. Inputs of the ASN are the position in the map and the Euclidean distance from the current position to the top-centre of the map, $< x, y, d >$, the output is the value of k (suitably scaled). The fitness function is the number of steps the controller needs to transverse the map and is limited to a maximum number of 1000 steps. Controllers exceeding this threshold are penalised with a fitness of 2000 steps. A population size of 500 is used.

5   Results

Results from controlling Chirikov's standard map using both ASN approaches are shown in Fig. 3. Both models led to effective controllers which were able to solve the problem (see Fig. 5(a)–(b)). The best performance comes from ASN_{mn}, but ASN_C can also lead to valid solutions when every pathway computes its dynamics independently or quasi-independently. The degree of crosstalk has a significant effect upon the solutions (see Fig. 4): low crosstalk seems to be beneficial, whereas high values add uncorrelated noise reducing the overall system behaviour. Similar conclusions on the effect of crosstalk were noted in [2].

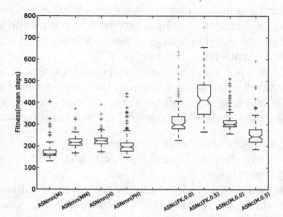

Fig. 3. State space targeting results using evolved ASNs with (M)ichaelis-Menten, (M)ulti-Dimensional (M)ichaelis-Menten, (H)ill, (P)robabilistic (H)ill, and (F)irst-order (K)inetics as regulatory equations. For ASN_C, the value next to the type of equation indicates the crosstalk rate. Summary statistics of the 100 runs are shown as box plots. Low values are better.

Perhaps the most interesting result is the capacity of ASN_C to solve the problem, even where there is no crosstalk. There is certainly an indication from the results that complex problems can be divided into smaller and independent tasks, which individually lead to valid solutions. A similar organization can be seen inside cells, which not only enclose a wide number of specific pathways, but also limit their interactions by using compartments. It highlights the essential role that crosstalk may play in the formation of more complex and realistic models. Despite the validity of the results, we believe that the procedure used to

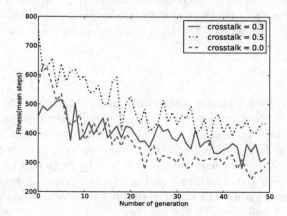

Fig. 4. Effect of crosstalk rate upon the effectiveness of ASN_C controllers, showing the rate of change in fitness 50 generations averaged over 100 runs

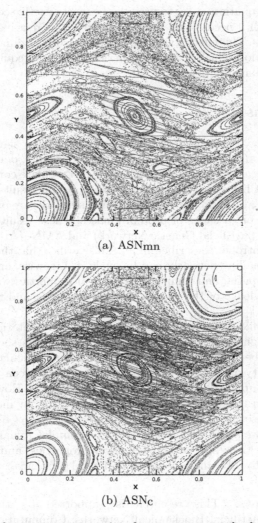

(a) ASN$_{mn}$

(b) ASN$_C$

Fig. 5. Example of state space targeting from an area at the bottom, $(0.45, 0) \rightarrow (0.55, 0.05)$, to a region at the top $(0.45, 0.95) \rightarrow (0.55, 1)$, in Chirikov's Standard Map with both ASN models. The ASN$_{mn}$ moves from the bottom to the top in 91 steps. The ASN$_C$ does it in 262 steps. The plotted standard map corresponds to the k value at the end of the ASN's execution.

determine the ASN$_C$ output (mean of all contributing networks) is not biologically plausible and therefore we might be losing some useful information. We hypothesise that ASN$_C$ would be better able to cope with incomplete or corrupt environmental information, enabling better environmental adaptation. This is something we aim to test in future work.

The choice of the regulatory function also has significant consequences. Generally, all regulatory functions work better for the ASN$_{mn}$ model. The

Michaelis-Menten equation seems to provide the most effective controllers in both approaches, however it needs to be complemented with a relatively high crosstalk probability in ASN_C. This effect contrasts with the results obtained using the first-order kinetics equation, which offer a lower performance and require a low crosstalk rate to achieve similar results (see Fig. 4).

6   Conclusions

In this paper we have presented an approach to modelling signalling pathways using evolutionary algorithms. Our results are encouraging; demonstrating that evolved artificial signalling networks can be used to regulate complex dynamical behaviours within Chirikov's standard map. Notably, our results show that effective controllers can also be found when signalling networks are interpreted as sets of either pathways or cascades of protein kinases. These results are broadly similar to the AGN- (Artificial Genetic Network) and AMN- (Artificial Metabolic Network) based controllers described in [9]. This verifies that the robustness and adaptability of signalling networks can be evolved. Likewise, an accurate representation of their spatial and temporal dynamical properties was achieved with no additional knowledge of the surrounding environment and the participating elements.

This paper has highlighted the importance that representation plays in effectively modelling signalling pathway dynamics. Our results show how different ASN models may be suited to different problems. In particular, we have illustrated the sensitivity of ASNs to the level of crosstalk between pathways, which has a large impact upon the effectiveness of the controllers. However, we believe crosstalk will be an important mechanism when looking for more complex and sophisticated representations of signalling networks.

In future work, we plan to explore how ASNs' complex dynamical behaviours can be affected by their spatial and temporal properties, and to look at how these can be used to solve more complex real-word problems.

Acknowledgements. This research is supported by the EPSRC ref: (EP/F060041/1), Artificial Biochemical Networks: Computational Models and Architectures.

References

1. Alon, U.: An introduction to systems biology: design principles of biological circuits. Chapman & Hall/CRC Mathematical and Computing Biology Series (2007)
2. Arias, M.A., Hayward, P.: Filtering transcriptional noise during development: concept and mechanisms. Nature Reviews Genetics 7(1), 34–44 (2006)
3. Bray, D., Lay, S.: Computer simulated evolution of a network of cell signalling molecules. Biophysical Journal 66, 972–977 (1994)
4. Chirikov, B.V.: Research concerning the theory of nonlinear resonance and stochasticity. Tech. rep., Institute of Nuclear Physics, Novosibirsk (1962)

5. Deckard, A., Sauro, M.H.: Preliminary studies on the *in silico* evolution of biochemical networks. Chembiochem. 5(10), 1423–1431 (2004)
6. Decraene, J., Mitchell, G.G., McMullin, B.: Evolving artificial cell signalling networks: Perspectives and Methods. SCI, vol. 6, pp. 167–186 (2009)
7. Huang, Z., Hahm, J.: Fuzzy modeling of signal transduction networks. In: Proc. 17th World Control, The International Federation of Automatic Control, pp. 15867–15872 (2008)
8. Klamt, S., Saez-Rodriguez, J., Lindquist, J.A., Simeoni, L., Giles, D.E.: A methodology for the structural and functional analysis of signalling and regulatory networks. BMC Bioinformatics 7(56), 1–26 (2006)
9. Lones, M.A., Tyrrell, A.M., Stepney, S., Caves, L.S.: Controlling Complex Dynamics with Artificial Biochemical Networks. In: Esparcia-Alcázar, A.I., Ekárt, A., Silva, S., Dignum, S., Uyar, A.Ş. (eds.) EuroGP 2010. LNCS, vol. 6021, pp. 159–170. Springer, Heidelberg (2010)
10. Morris, K.M., Saez-Rodriguez, J., Sorger, K.P., Lauffenburger, A.D.: Logic-based models for analysis of cell signalling networks. Biochemistry 4(49), 3216–3224 (2010)
11. Sachs, K., Gifford, D., Jaakkola, T., Sorger, P., Lauffenburger, D.A.: Bayesian networks approach to the cell signalling pathway modeling. Science's STKE 148, 38–42 (2002)
12. Stepney, S.: Nonclassical computation: a dynamical systems perspective. In: Rozenberg, G., Bäck, T., Kok, N.J. (eds.) Handbook of Natural Computing, vol. 2, ch. 52. Springer, Heidelberg (2011)
13. Said, M.R., Oppenheim, A.V., Lauffenburger, D.A.: Modelling cellular signal processing using interacting Markov chains. In: Proc. International Conference on Acoustic, Speech, Signal Processing (ICASSP 2003), Hong Kong, pp. 41–44 (2003)
14. Schroer, C.G., Ott, E.: Targeting in Hamiltonian systems that have mixed regular/chaotic phase spaces. Chaos 7, 512–519 (1997)
15. Tulupyev, A., Nikolenko, S.: Directed Cycles in Bayesian Belief Networks: Probabilistic Semantics and Consistency Checking Complexity. In: Gelbukh, A., de Albornoz, Á., Terashima-Marín, H. (eds.) MICAI 2005. LNCS (LNAI), vol. 3789, pp. 214–223. Springer, Heidelberg (2005)
16. Ziegler, J., Banzhaf, W.: Evolving control metabolisms for a robot. Artificial Life 7, 171–190 (2001)

The Effect of Membrane Receptor Clustering on Spatio-temporal Cell Signalling Dynamics

Bertrand R. Caré[1,3] and Hédi A. Soula[2,3]

[1] Université de Lyon,
Laboratoire d'InfoRmatique en Image et Systèmes d'information, CNRS UMR5205,
F-69621 VILLEURBANNE
bertrand.care@insa-lyon.fr
[2] Université de Lyon,
Cardiovasculaire, Métabolisme, Diabétologie et Nutrition, Inserm UMR1060,
F-69621 VILLEURBANNE
hedi.soula@insa-lyon.fr
[3] EPI BEAGLE INRIA

Abstract. Membrane receptors allow the cell to respond to changes in the composition of its external medium. The ligand-receptor interaction is the core of the signalling process and may be greatly influenced by the spatial configuration of receptors. As growing pieces of evidence suggest that receptors are not homogeneously spread on the cell surface, but tend to form clusters, we propose to investigate the implication of receptor clustering on ligand binding kinetics using a computational individual-based model. The model simulates the activation of receptors distributed in clusters or uniformly spread. The tracking of binding events allows the analysis of the effect of receptor clustering through the autocorrelation of the receptor activation signal and the empirical time distributions of binding events, which are still unreachable with in vitro or in vivo experiments. Results show that the apparent affinity of clustered receptors is decreased. Additionally, receptor occupation becomes spatially and temporally correlated, as clustering creates platforms of coherently activated receptors. Changes in the spatial characteristics of a signalling system at the microscopic scale globally affect its function in time and space.

Keywords: cell signalling, receptor, ligand, clustering, pathway, binding, kinetics, equilibrium, autocorrelation, individual-based model, computational biology.

1 Introduction

In cell signalling, most models describe the ligand as an external stimulus and the receptor as the binding target, based on the ground of chemical enzyme/substrate formalism [1, 2]. Such formulations are based on the law of mass-action, which evaluates local reaction rates from averaged chemical species densities over the medium volume. The law of mass-action is a mean-field approximation since it

M.A. Lones et al. (Eds.): IPCAT 2012, LNCS 7223, pp. 50–61, 2012.
© Springer-Verlag Berlin Heidelberg 2012

estimates local reaction rates on the basis of average values of the reactants densities over a large spatial domain. In addition, it amounts to assume that ligand-receptor interactions are independent with respect to time and space [3, 4].

In biology, these assumptions can be questioned, in particular when considering membrane receptors which are restricted to only 2 of the 3 spatial dimensions [5, 6]. On the specific case of membrane-restricted receptors (on spherical cells), the expression for reaction rate coefficients is a non-linear function of cell surface receptor density [7]. This pioneer study has been enriched by further works towards reversibility and rebinding, [8], receptor density [9], time dependency [10], and gradient sensing capabilities [11, 12].

Furthermore, the spatial organization of receptors *on the membrane itself* should also be taken into account. At first glance, since membrane receptors are bound to the cell membrane that allows for lateral degrees of freedom, one would expect a simple (and homogeneous) distribution of receptors on the membrane. Indeed, cell membrane is composed of a mixture of phospholipids in a fluid phase and as such, in the classical fluid-mosaic model of membrane [13], membranes components undergo isotropic random movement akin to Brownian motion [14, 15]. In this model, the resulting equilibrium distribution of components – and receptors among them – is therefore homogeneous. Recently, however, this picture has evolved considerably towards a non-homogeneous distribution of the usual components of cell membranes [16, 17, 18, 19, 20]. More and more evidence points towards the existence of micro-domains enriched in various lipids, such as cholesterol, as well as other proteins, such as receptors. In particular, receptor colocalization in lipid rafts and other membrane structures have been reported [21, 22, 23]. This specific localization and clustering may have a dramatic influence on signalling. This influence however remains unclear as literature reports contradictory effects of clustering/declustering on signalling (see e.g. [24, 25, 22]). The method used to disrupt the clusters of receptors may have significant side-effects on the cell signalling system.

On the modelling side, the impact of an inhomogeneous receptor density *on* the membrane itself has been studied only recently. Only few theoretical contributions have been reported in some specific cases : bacteria sensitivity[26] and chemotaxis [27], G-protein activation [28], simple model of trans-phosphorylation (implying two receptors only) [29]. In addition, several more detailed studies illustrate the possible effect of receptor clustering on receptor binding by inducing enhanced rebinding and ligand receptor switching [30, 31, 32, 33], or enhancing encounter probability of activated receptors with submembranar signalling proteins such as in GPCR signalling pathways [34]. Notably [32] proposes that clustering provides higher rebinding capabilities and therefore helps to obtain a better response – i.e. more binding events. However, another analysis [8] proposes that the forward rate constant is diminished when receptors are clustered, providing in that case less binding events. Both effects counteract themselves, and the final output remains to be studied.

In a previous paper, we investigated how receptor distribution may impact the primordial step of signalling that is ligand binding to receptor extracellular domain [35]. We showed that in the case of a diffusion-limited reaction, receptor clustering impairs the sensitivity of the signalling system. While conserving the microscopic binding properties, the apparent affinity of a receptor for its ligand diminishes with clustering. We showed that this effect is based on spatial features and is diffusion-dependent. In the limit of high diffusion this impairment vanishes, whereas low diffusion amplifies it.

We present in this article a detailed study on how this effect takes place in terms of binding. Intuitively two effects are in action. Clustered receptors are "harder to find", as it diminishes their probability to be found by ligand molecules. In the other hand, when receptors are clustered, they are more likely to be found by a ligand that has been released by another nearby receptor. In other words, more rebinding events are expected in the clustered case. Obviously these two effects counter themselves and the outcome is not intuitively clear. Additionnaly, we show in this article several properties of the binding kinetics of receptors depending on their spatial configuration. Especially, we investigated not only how clustering affects the global amount of activation resulting from ligand stimulation, but also how the temporal dynamics of receptor activation changes with clustering, which translates a spatial correlation into a temporal one.

2   Models

As already mentioned, mathematical models of binding kinetics generally rely on the law of mass action. In the case of a correlated receptor spatial configuration, this hypothesis breaks down. In order to investigate this issue, we developped a simulation engine where the spatial characteristics of real signalling systems arises naturally by using an individual-based model. This simulation engine is defined and described in detail in another article [35] that we briefly describe here as well. The engine computes the equation of movement of punctual particles in a 2-dimension space with cylindric boundary conditions on the x-axis, and closed boundary on the y-axis, the membrane being at $y = 0$. This space is used to describe the extracellular medium. The membrane is the bottom line of the 2-dimension space. Receptors are positioned on the membrane and do not move during simulation, assuming that receptor diffusion is negligible compared to ligand diffusion. Ligand molecules are punctual particles which undergo a classical 2-dimension Brownian motion in the extracellular space. As mentioned above, motion is forbidden beneath the membrane or through the upper part of the simulation space. However, particles going through one lateral boundary appear across the other. Ligand molecules undergo Brownian motion in the overdamped regime via an explicit Euler scheme of step dt :

$$x(t + dt) = x(t) + \sqrt{Ddt}Z_1$$
$$y(t + dt) = y(t) + \sqrt{Ddt}Z_2$$

with D being the simulated ligand diffusion coefficient, and $Z_{1,2}$ are random values drawn from a normalized Gaussian variate. Binding can occur whenever a ligand molecule is in the 'affinity zone' of a unoccupied receptor – a fixed square above the position of the receptor. If the receptor is free – not already bound to a ligand – binding can occur with a given probability p_1. Finally, an already bound ligand molecule can be released at the border of the affinity zone with a probability p_1 at each time step.

We studied two kinds of receptor spatial configurations in these simulations. The first is a reference – control – receptor distribution, in which they are uniformly spread on the 1-dimension membrane – referred hereafter as to the homogeneous distribution, or unclustered receptors case. The clustered case is obtained by positioning receptors next to each other – with adjacent but non-overlapping affinity zones – by groups of n. These clusters are then uniformly spaced. Most simulations will then compare several cluster sizes (various n) to the control. Note that the control case describes this reaction :

$$L + R \underset{k_{-1}}{\overset{k_1}{\rightleftharpoons}} C \tag{1}$$

and [35] showed that we can relate reaction rates to the binding/unbinding probabilities via :

$$k_{-1} = p_{-1}$$
$$k_1 = \frac{p_1 S_r}{S_t}$$

with S_r being the area of the affinity zone and S_t the total area of the extracellular medium.

3   Results

Unless stated otherwise, the number of receptors for each simulation run was $N_r = 500$, the number of ligand molecules $N_l = 4.10^5$, $k_1 = k_{-1} = 10.0$, $dt = 10^{-3}$ giving $p_1 = p_{-1} = 10^{-2}$. The surface of each affinity zone was $S_r = 0.4$ and the total medium surface $S_t = 2.10^5$. The ligand diffusion coefficient was $D = 1.0$. The cluster size is noted n, $n = 1$ referring to the case of homogeneously spread receptors.

3.1   Transient Phase

Our previous results only dealt with receptor occupation at equilibrium, i.e. the average number of ligand-receptor complexes after the simulation reached a stationay state. The transient solution should yield the same result : clustering decreases the overall responses. As shown in Fig. 1 the fraction of occupied receptors through time was also cluster-dependent. The figures show a similar initial activation rise. Indeed, initially, ligand molecules were positioned uniformly, and since the global surface covered by receptor affinity zones was unchanged by clustering, the initial probability for a ligand to be in an unoccupied receptor was equal no matter the cluster size. Quickly afterwards though, binding events began to decline steadily whenever receptor were clustered. This shows that the actual binding history for ligand molecules in the vicinity of receptor must be taken into account in order to understand this shift in complexation.

3.2   Binding Events Analysis

The occurence of specific events was tracked during simulation runs. The simulation yielded simultaneously the number of binding events and the number of ligand-receptor encounter events that took place at each time step. Binding events fell into two categories: the first binding events and the rebinding events. The former refers to ligand molecules binding to a receptor for the first time, from the ligand point of view. The latter refers to ligand molecules binding to a receptor for at least the second time, from the ligand point of view.

The relative contribution of binding events of each kind versus cluster size is reported on Fig. 2. In order to avoid any bias due to the decreasing in receptor occupation with clustering, the number of events were normalized on the total number of binding events recorded. As clustering increases, the contribution of first binding events dropped dramatically, while the amount of receptor activation due to rebinding increased. First binding events occured less often if receptors were clustered, but clustering was favorable to rebinding. This suggests that most of the receptor activation was performed by a small contingent of ligand that kept on rebinding.

By computing the ratio of the number of rebinding events to the number of first binding events versus cluster size (see Fig. 3), we obtained the average number of times a ligand molecule rebound to a receptor. As expected this ratio increased with cluster size. By having access to the index of each ligand molecule that generated a binding event, we also obtained the number of unique ligand molecules that had contributed to the total number of binding events. This gives an estimate of the average number of binding events generated by a single ligand molecule according to the cluster size – Fig. 3. Both curves have a similar trend : in the clustered case, receptor activation was induced through constant rebinding by the same set of ligands. Indeed, a high number of unique rebinding indicates a small contingent of ligand molecules involved in the signal. This put a strong emphasis on dependence on the binding history of ligands. On the other hand, in the unclustered case, most binding was performed by 'fresh' ligands newly coming from the medium, whereas rebinding was marginal.

3.3   Ligand Temporal Dynamics

The simulation also provided the time a ligand molecule had to wait between two consecutive binding events. Here, "consecutive" is defined in the ligand molecule referential. Consecutive binding events, that is, rebinding events, were sorted out in two classes : rebinding by a ligand molecule to the same receptor (self-rebinding) and rebinding by a ligand molecule to a different receptor (distinct rebinding). It was thus possible to investigate the qualitative effects of receptor

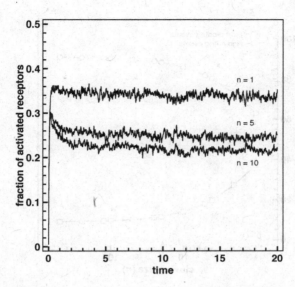

Fig. 1. Fraction of activated receptors versus time for various cluster sizes. The graph shows the signals of receptor activation for a single simulation run with the same parameters except the receptor clusters size. Clustering decreased the receptor activation at equilibrium.

clustering on the temporal dynamics of binding. Fig. 4 shows the mean time between rebinding events sorted in the two types mentioned above, plus the mean time of all rebinding times indifferently, for different cluster sizes. As expected, the time to rebind to another receptor decreased with clustering - since there were other receptors available in the vicinity when they were clustered. In the unclustered case, rebinding to another receptor was a marginal event, as suggested by its longer mean time (one order of magnitude above the others) and its quasi-inexistent influence on the overall rebinding time. Additionally, we noted that the self rebinding time also decreased with clustering, making the self-rebinding more frequent in the clustered case. This could be explained by the fact that, in the unclustered case, a bound receptor could be readily reoccupied by a new ligand molecule. We also had access to inter and intra-cluster rebinding times. Inter-cluster rebinding refers to rebinding of a ligand molecule to a receptor belonging to another cluster, unlike intra-cluster rebinding where rebinding occur to a receptor of the same cluster. Simply put, in the clustered case, there were no rebinding events (during simulation time) between clusters. All rebinding occured within the same cluster. As for the unclustered case, each receptor is a single cluster and we already mentioned that inter-cluster rebinding was extremely marginal.

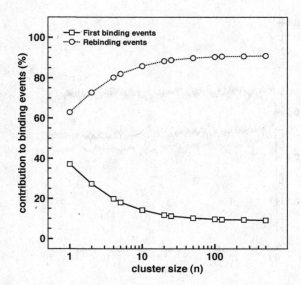

Fig. 2. Relative contribution of binding events from different types to the total binding. Binding events were splitted in two distinct types : the first binding type, i.e. when a ligand molecules bound to a receptor for the first time, and the rebinding type, i.e. when a ligand bound to a receptor and had already been bound in the past to any receptor.

3.4   Receptor Temporal Dynamics

From a receptor point of view, the change in the temporal dynamics of rebinding suggests that the spatial correlation of positions should induce a temporal correlation of activation. In order to investigate this coupling, the activation signal of each receptor was tracked for each time step in the form of a binary signal (0 : free, 1 : occupied by ligand). This signal was then averaged for 10 neighboring receptors. For all n, it simply means we sorted by groups of the 10 closest receptors. The autocorrelations of such signals were computed and are compared in Fig. 5 (dashed lines) with the autocorrelation of a spatially uncorrelated signal (solid line, $n = 1$). The autocorrelation is the correlation of the signal with itself shifted by a lag. Let $x(t)$ being a signal, we simply computed the following expression, the average being taken over the entire time course :

$$ac(lag) = \langle (x(t) - \bar{x}) \, (x(t + lag) - \bar{x}) \rangle$$

The theoretical autocorrelation for binding events was expected to follow an exponential decay. Indeed, the curve for $n = 1$ presented a classical exponential decay. The correlation of the activation signal decreased with time. However, as clustering was introduced, the half-time of this decay increased. This means that the activation state of receptors correlated with their past state for a longer time

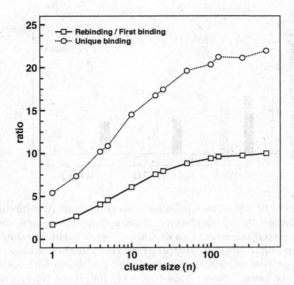

Fig. 3. Squares, solid line : ratio of rebinding events to first binding events (squares, solid line) versus cluster size. A ratio of 5 indicates that, in average, 5 out of 6 binding events occured through rebinding. Circles, dashed line : ratio of the number of individual ligands involved in binding events to the total number of binding events versus cluster size. In this case, a ratio of 5 means that, over the course of the simulation, a unique ligand molecule generated in average 5 binding events on its own (ignoring the ligand molecules that never bound to a receptor).

with clustering than in the unclustered case. The autocorrelation profiles suggest that the temporal correlation of the activation state of adjacent receptors was stronger with clustering.

Clustering introduced a spatial correlation on receptor activation, shown by an increase in rebinding events at the expense of first binding events. Globally, the fraction of activated receptors, at equal ligand stimulation, was decreased, as rebinding did not overcome the loss of encounter events between ligand molecules and receptor. The effect of clustering also appeard on the temporal dynamics of the receptor system, as the activation state of receptors correlated more with its past value. This illustrates how the spatial correlation of receptors translates into a temporal correlation of their binding with the ligand.

4   Discussion

A computational model was used to recreate ligand-receptor binding under specific spatial configurations similar to the ones observed experimentally. This kind of model allows for a detailed analysis of signalling systems, as each individual binding event can be tracked.

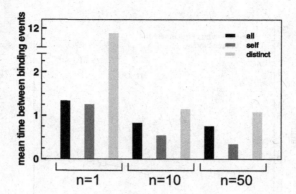

Fig. 4. The time spent by ligand molecules between two consecutive binding events was saved for each molecule during simulations. Theses durations were sorted out in two types : the times between rebinding to a distinct receptor (distinct rebinding) and the times between rebinding to the same receptor (self-rebinding). The mean time between consecutive binding events of such kinds (greyscales) are shown with respect to cluster size n, along with the mean time of consecutive binding when both types are pooled (black).

Receptor clustering seemingly induced a quantitative effect that decreased the global receptor activation by an external ligand. The behavior of the simulated signalling system could be examined in depth. When clustering was imposed to receptors, ligand binding occured more because of ligand molecules rebinding to receptors, at the expense of ligand molecules finding and binding for the first time to a receptor. Not only the crude number of such events was altered in favor of rebinding, the time spent between consecutive binding events also changed. The activation signal of receptors becomes space and time-dependent, showing how a different receptor spatial configuration introduced a shift in the temporal dynamics of the signal transmitted.

This suggests that the peculiar spatial distributions of receptors observed in nature might have a functional role in signalling. This role could possibly be not only quantitative, as the global receptor activation is reduced with clustering, but also qualitative. This study suggests that clustering introduces platforms of aggregated receptors whose activation becomes correlated in time and space, that is, the correlation of receptor position translates into a synchronization of receptor activation. This property is not available in the homogeneous receptor repartition scenario, where receptors are activated randomly in space and time. Making the activation of receptors time and space-dependent could be an advantage in terms of sensitivy, noise reduction and signal robustness. It could also improve signalling-associated cellular processes such as receptor trafficking, recycling, and interaction between parallel pathways. For instance, ligand-induced receptor internalization was observed in different pathways [36, 37], and could be partially relying on harmonization of receptor activation achieved by clustering : activated receptors can be internalized and recycled more efficiently if they

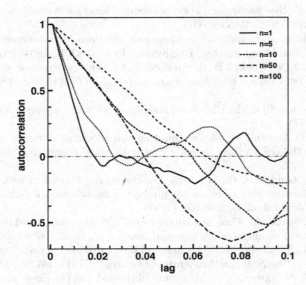

Fig. 5. Autocorrelation functions of the receptor activation signal for various cluster sizes. The binary occupation signal was computed for each receptor and each time step. The average signal of 10 neighboring receptors was used to perform an autocorrelation computation. Autocorrelation functions for clustered receptors show a longer exponential decay, suggesting that the spatial correlation between receptors translated into a temporal correlation.

are already grouped together, rather than spread ramdomly on the cell surface. The question remains to be investigated in studies integrating the spatial and temporal characteristics of such processes, using both modelling and biological experiments.

Acknowledgments. BC holds a fellowship from la Région Rhône-Alpes. We gratefully acknowledge support from the CNRS/IN2P3 Computing Center (Lyon / Villeurbanne, France), for providing a significant amount of the computing resources needed for this work.

References

[1] Heffetz, D., Zick, Y.: Receptor aggregation is necessary for activation of the soluble insulin receptor kinase. The Journal of Biological Chemistry 261(2), 889–894 (1986)
[2] Flrke, R.R., Schnaith, K., Passlack, W., Wichert, M., Kuehn, L., Fabry, M., Federwisch, M., Reinauer, H.: Hormone-triggered conformational changes within the insulin-receptor ectodomain: requirement for transmembrane anchors
[3] Murray, J.D.: Mathematical Biology: I. An Introduction. Springer, Heidelberg (2002)

[4] Gillespie, D.T.: Stochastic simulation of chemical kinetics. Annual Review of Physical Chemistry 58(1), 35–55 (2007)

[5] Berry, H.: Monte carlo simulations of enzyme reactions in two dimensions: fractal kinetics and spatial segregation. Biophysical Journal 83(4), 1891–1901 (2002)

[6] Kholodenko, B.N., Hoek, J.B., Westerhoff, H.V.: Why cytoplasmic signalling proteins should be recruited to cell membranes. Trends in Cell Biology 10(5), 173–178 (2000)

[7] Berg, H.C., Purcell, E.M.: Physics of chemoreception. Biophysical Journal 20(2), 193–219 (1977)

[8] Goldstein, B., Dembo, M.: Approximating the effects of diffusion on reversible reactions at the cell surface: ligand-receptor kinetics. Biophysical Journal 68(4), 1222–1230 (1995)

[9] Erickson, J., Goldstein, B., Holowka, D., Baird, B.: The effect of receptor density on the forward rate constant for binding of ligands to cell surface receptors. Biophysical Journal 52(4), 657–662 (1987)

[10] Zwanzig, R., Szabo, A.: Time dependent rate of diffusion-influenced ligand binding to receptors on cell surfaces. Biophysical Journal 60(3), 671–678 (1991)

[11] Endres, R.G., Wingreen, N.S.: Accuracy of direct gradient sensing by single cells. Proceedings of the National Academy of Sciences 105(41), 15749–15754 (2008)

[12] Endres, R.G., Wingreen, N.S.: Maximum likelihood and the single receptor. Physical Review Letters 103(15), 158101 (2009); PMID: 19905667

[13] Singer, S.J., Nicolson, G.L.: The fluid mosaic model of the structure of cell membranes. Science 175(23), 720–731 (1972)

[14] Koppel, D.E., Sheetz, M.P., Schindler, M.: Matrix control of protein diffusion in biological membranes. Proceedings of the National Academy of Sciences of the United States of America 78(6), 3576–3580 (1981)

[15] Chung, I., Akita, R., Vandlen, R., Toomre, D., Schlessinger, J., Mellman, I.: Spatial control of EGF receptor activation by reversible dimerization on living cells. Nature 464(7289), 783–787 (2010)

[16] Simons, K., Ikonen, E.: Functional rafts in cell membranes. Nature 387(6633), 569–572 (1997)

[17] Simons, K., Toomre, D.: Lipid rafts and signal transduction. Nature Reviews. Molecular Cell Biology 1(1), 31–39 (2000)

[18] Schuck, S., Simons, K.: Polarized sorting in epithelial cells: raft clustering and the biogenesis of the apical membrane. Journal of Cell Science 117(25), 5955–5964 (2004)

[19] Brown, D.A., London, E.: Functions of lipid rafts in biological membranes. Annual Review of Cell and Developmental Biology 14(1), 111–136 (1998)

[20] Zhang, J., Leiderman, K., Pfeiffer, J.R., Wilson, B.S., Oliver, J.M., Steinberg, S.L.: Characterizing the topography of membrane receptors and signaling molecules from spatial patterns obtained using nanometer-scale electron-dense probes and electron microscopy. Micron. 37(1), 14–34 (2006) (Oxford, England: 1993)

[21] Gustavsson, J., Parpal, S., Karlsson, M., Ramsing, C., Thorn, H., Borg, M., Lindroth, M., Peterson, K.H., Magnusson, K.-E., Strälfors, P.: Localization of the insulin receptor in caveolae of adipocyte plasma membrane. The FASEB Journal 13(14), 1961–1971 (1999)

[22] Parpal, S.: Cholesterol depletion disrupts caveolae and insulin receptor signaling for metabolic control via insulin receptor substrate-1, but not for mitogen-activated protein kinase control. Journal of Biological Chemistry 276(13), 9670–9678 (2000)

[23] Lee, S., Mandic, J., Van Vliet, K.J.: Chemomechanical mapping of ligandreceptor binding kinetics on cells. Proceedings of the National Academy of Sciences of the United States of America 104(23), 9609–9614 (2007)

[24] Lim, K., Yin, J.: Localization of receptors in lipid rafts can inhibit signal transduction. Biotechnology and Bioengineering 90(6), 694–702 (2005)

[25] Vitte, J., Benoliel, A.-M., Eymeric, P., Bongrand, P., Pierres, A.: Beta-1 integrin-mediated adhesion be initiated by multiple incomplete bonds, thus accounting for the functional importance of receptor clustering. Biophysical Journal 86(6), 4059–4074 (2004)

[26] Bray, D., Levin, M.D., Morton-Firth, C.J.: Receptor clustering as a cellular mechanism to control sensitivity. Nature 393, 85–88 (1998)

[27] Mello, B.A., Shaw, L., Tu, Y.: Effects of receptor interaction in bacterial chemotaxis. Biophysical Journal 87(3), 1578–1595 (2004)

[28] Mahama, P.A., Linderman, J.J.: A monte carlo study of the dynamics of g-protein activation. Biophysical Journal 67(3), 1345–1357 (1994)

[29] Wanant, S., Quon, M.J.: Insulin receptor binding kinetics: modeling and simulation studies. Journal of Theoretical Biology 205(3), 355–364 (2000)

[30] Shea, L.D., Omann, G.M., Linderman, J.J.: Calculation of diffusion-limited kinetics for the reactions in collision coupling and receptor cross-linking. Biophysical Journal 73(6), 2949–2959 (1997)

[31] Shea, L.D., Linderman, J.J.: Compartmentalization of receptors and enzymes affects activation for a collision coupling mechanism. Journal of Theoretical Biology 191(3), 249–258 (1998)

[32] Gopalakrishnan, M.: Effects of receptor clustering on ligand dissociation kinetics: Theory and simulations. Biophysical Journal 89(6), 3686–3700 (2005)

[33] Ghosh, S., Gopalakrishnan, M., Forsten-Williams, K.: Self-consistent theory of reversible ligand binding to a spherical cell. Physical Biology 4(4), 344–354 (2008)

[34] Fallahi-Sichani, M., Linderman, J.J.: Lipid Raft-Mediated regulation of G-Protein coupled receptor signaling by ligands which influence receptor dimerization: A computational study. PLoS ONE 4(8), e6604 (2009)

[35] Caré, B.R., Soula, H.A.: Impact of receptor clustering on ligand binding. BMC Systems Biology 5(1), 48 (2011)

[36] Carpentier, J.L., Paccaud, J.P., Gorden, P., Rutter, W.J., Orci, L.: Insulin-induced surface redistribution regulates internalization of the insulin receptor and requires its autophosphorylation. Proceedings of the National Academy of Sciences of the United States of America 89(1), 162–166 (1992)

[37] Giudice, J., Leskow, F.C., Arndt-Jovin, D.J., Jovin, T.M., Jares-Erijman, E.A.: Differential endocytosis and signaling dynamics of insulin receptor variants IR-A and IR-B. Journal of Cell Science 124(Pt 5), 801–811 (2011)

Systems Biology Analysis of Kinase Inhibitor Protein Target Profiles in Leukemia Treatments

Jacques Colinge[*], Uwe Rix, Keiryn L. Bennett, and Giulio Superti-Furga

Center for Molecular Medicine of the Austrian Academy of Sciences
AKH-BT 25.3, Lazarettgasse 14, Vienna, Austria
jcolinge@cemm.oeaw.ac.at

To be able to understand the mechanisms of action of drugs, predict their efficacy, and anticipate their potential side-effects is important during drug development. In diseases where the genetic background of patients modulates treatment response, it might allow personalizing the therapy.

Substantial progress in proteomic technologies[1] have made it possible to develop chemical proteomics methods, where the protein targets of a drug are affinity-purified and identified by mass spectrometry[2, 3]. Compound-protein interactions are measured in a biological context as opposed to *in vitro* binding assays. That is, drug-protein interactions can not only be determined proteome-wide, but also in a tissue- or cell type-dependent manner.

Drugs induce global perturbations of treated cells[4]. Targeted proteins are involved in biochemical reactions that take place within one or several biological pathways, which can be in interaction with other pathways[5]. Consequently, to act on a single protein activity that is part of a complex network can have far-reaching consequences on a multitude of biological functions. Moreover, compounds often have more than one target thus making the effects of their administration even broader. For instance, the tyrosine kinase inhibitor imatinib, which is a hallmark of targeted therapy against the chronic myeloid leukemia-causing fusion protein BCR-ABL, was found to have at least five additional potent targets, including the non-kinase (NQO2)[6].

To reach the promises of chemical proteomics we have developed computational methods to analyze unbiased drug target profiles, which we applied to kinase inhibitors that are used in cancer treatment, leukemia mainly. Kinase inhibitors tend to have large and complex target spectra and they therefore perturb the regulation of phosphorylation-based signaling pathways deeply and broadly. They constitute a perfect model case to investigate global drug-induced changes of the molecular biology of a cell and we naturally decided to follow a systems biology approach o this problem. We map the drug protein targets on the human interactome, which is provided by the integration of protein-protein physical interactions found in several public databases, and by means of diffusion methods[7] we score the entire network for association strength with the drug action to obtain a treatment model[8, 9]

---

[*] Corresponding author.

M.A. Lones et al. (Eds.): IPCAT 2012, LNCS 7223, pp. 62–66, 2012.
© Springer-Verlag Berlin Heidelberg 2012

(Figure 1A). This procedure can exploit an estimation of the individual drug target affinities available from the proteomics data, e.g. spectral counts, to weigh the target contributions to the treatment model. Similarly, known genes or proteins at the source of the disease can be mapped on the network and a disease model obtained. By considering the overlap of the two models and the disease or treatment association scores, we can determine a drug treatment efficacy score.

We have shown in the case of chronic myeloid leukemia (CML) that the efficacy scores of 4 kinase inhibitors, imatinib, dasatinib, bosutinib, and bafetinib, reflected their relative efficacy in the clinical practice[8] (bosutinib still in development) and ranked CML as a very likely successful application area in comparison with other tumors and diseases (Figure 1B), which confirmed our scoring approach. We have also demonstrated that a patient specific disease model including a classical imatinib-resistance gene (LYN) resulted in an increase of dasatinib, bosutinib, and bafetinib, which are second-generation compounds especially designed to target SRC kinases additionally, including LYN. Such a score increase did not happen for imatinib. This indicated that patient profiles can be exploited to determine appropriate therapeutic options on a personal basis, combining the target spectrum knowledge and patient data appropriately. By further analyzing the ranks of various tumors and other diseases against which we scored the 4 inhibitors, we could for instance propose that dasatinib should be tested against the lung cancer and hepataocellular carcinoma. Dasatinib lung cancer efficacy was confirmed by another study we published[10]. We also detected a potential effect of dasatinib, bafetinib, and bosutinib against the Noonan syndrome, which is plausible as this syndrome is related to kinases (RAS) and kinase inhibitors are tested as therapies[11].

We also examined the bosutinib target profile by considering its treatment model on the interactome only (no disease) and by scoring KEGG pathways with the treatment association scores. It naturally ranked the CML pathway on the first place but also retrieved several immune system-related pathways in the top 5% significant pathway list. Detailed inspection of the area of the interactome that is most impacted by bosutinib treatment (5% significance), revealed that numerous targets and their interactors are described in the literature for causing immunosuppression upon deactivation, e.g. by a kinase inhibitor. This prediction is additionally supported by documented immunosuppressive effects of dasatinib, which is already in clinical use, whereas bosutinib is still in development. This last CML result confirms the potential of our approach to infer likely side effects, which can be crucial to address safety issues early on or, in the perspective of drug development economics, to stop dangerous compounds before excessive development costs are invested.

In a recent study (Rix et al., in revision), we profiled dasatinib, bosutinib, nilotinib, and bafetinib in Philadelphia positive acute lymphoblastic leukemia (Ph+ ALL) cell lines and we compared the computed drug treatment efficacy scores with actual IC_{50} measures. The experimental results matched the prediction well with dasatinib being the most effective compound, followed by bafetinib and nilotinib that had medium

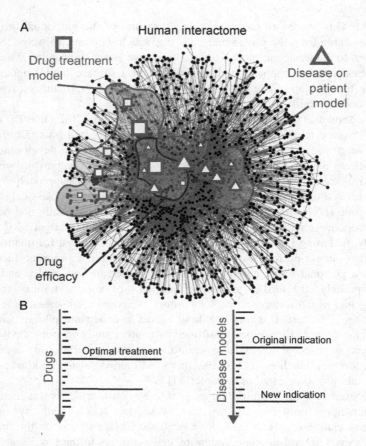

Fig. 1. Principle of scoring drug treatment efficacy. (A) Drug protein targets (squares) are mapped on the human interactome and an area of significant influence ("drug treatment model") is determined by means of diffusion methods and randomization. The same procedure is applied to genes and proteins known to be the cause of the disease of interest (triangles). Where the two areas of significant influence overlap a correlation score can be computed to evaluate the likelihood of a specific drug treatment on a given disease. (B) The ability of scoring drug treatment efficacies can be exploited in two dual ways: to infer the adequacy of a drug treatment for a specific patient by integrating its genetic background in a dedicated disease model; or to infer repurposing of compounds by comparing a specific compound against several disease models.

efficacy, and by bosutinib that had little effect though it is a potent BCR-ABL inhibitor and it had the largest target spectrum. In this new work we used both an average patient model, considering the relative frequencies of gene deletions in Ph+ ALL patients[12], and detailed genetic maps provided by the COSMIC and Oncomine databases. Encouraged by this accurate prediction, we performed similar measurements upon LYN knock-down, LYN being a prominent dasatinib target and a target of the other 3 inhibitors as well. Again, the in silico estimation of drug efficacies obtained by removing LYN from the interactome were in remarkable adequacy with the experiments since we predicted dasatinib to be the only strongly perturbed drug, with a reduction of its action, what the new IC_{50}'s confirmed.

Altogether, our results with kinase inhibitors and CML and Ph+ ALL indicate that the precise knowledge of drug target spectra as measured by chemical proteomics can be integrated with global models of cell biology such a protein interactions or pathways and knowledge on disease causes to obtain important predictions of efficacy and side effects. Moreover, considering multiple diseases, existing compounds can be repurposed. In this work, no specific leukemia-related characteristics were exploited and it is very likely that the presented computational methods can be applied to a broad range of pathologies and compound classes. In general, it shows that the determination of complete and accurate – tissue-type dependent – drug profiles have a great potential to synergize with DNA deep sequencing technologies to establish effective therapeutic programs informed by individual genetic profiles. It also demonstrates that global signaling perturbation can be properly captured at macroscopic scale through the integration of protein physical interaction maps.

References

1. Domon, B., Aebersold, R.: Mass spectrometry and protein analysis. Science 312, 212–217 (2006)
2. Bantscheff, M., Scholten, A., Heck, A.J.: Revealing promiscuous drug-target interactions by chemical proteomics. Drug Discov. Today 14, 1021–1029 (2009)
3. Rix, U., Superti-Furga, G.: Target profiling of small molecules by chemical proteomics. Nat. Chem. Biol. 5, 616–624 (2009)
4. Araujo, R.P., Liotta, L.A., Petricoin, E.F.: Proteins, drug targets and the mechanisms they control: the simple truth about complex networks. Nat. Rev. Drug Discov. 6, 871–880 (2007)
5. Keith, C.T., Borisy, A.A., Stockwell, B.R.: Multicomponent therapeutics for networked systems. Nat. Rev. Drug Discov. 4, 71–78 (2005)
6. Rix, U., Hantschel, O., Durnberger, G., Remsing Rix, L.L., Planyavsky, M., Fernbach, N.V., Kaupe, I., Bennett, K.L., Valent, P., Colinge, J., Kocher, T., Superti-Furga, G.: Chemical proteomic profiles of the BCR-ABL inhibitors imatinib, nilotinib, and dasatinib reveal novel kinase and nonkinase targets. Blood 110, 4055–4063 (2007)
7. Kohler, S., Bauer, S., Horn, D., Robinson, P.N.: Walking the interactome for prioritization of candidate disease genes. Am. J. Hum. Genet. 82, 949–958 (2008)
8. Colinge, J., Rix, U., Superti-Furga, G.: Novel global network scores to analyze kinase inhibitor profiles. In: Chen, L., Zhang, X., Shen, B., Wu, L., Wang, Y. (eds.) 4th International Conference on Computational Systems Biology, vol. 13, pp. 305–313. World Publishing Company, Suzhou (2010)
9. Colinge, J., Rix, U., Bennett, K.L., Superti-Furga, G.: Systems biology analysis of protein-drug interactions. Proteomics Clin. Appl. (in press)
10. Li, J., Rix, U., Fang, B., Bai, Y., Edwards, A., Colinge, J., Bennett, K.L., Gao, J., Song, L., Eschrich, S., Superti-Furga, G., Koomen, J., Haura, E.B.: A chemical and phosphoproteomic characterization of dasatinib action in lung cancer. Nat. Chem. Biol. 6, 291–299 (2010)

11. Gelb, B.D., Tartaglia, M.: Noonan syndrome and related disorders: dysregulated RAS-mitogen activated protein kinase signal transduction. Hum. Mol. Genet. 15 Spec No 2, R220-226 (2006)
12. Mullighan, C.G., Miller, C.B., Radtke, I., Phillips, L.A., Dalton, J., Ma, J., White, D., Hughes, T.P., Le Beau, M.M., Pui, C.H., Relling, M.V., Shurtleff, S.A., Downing, J.R.: BCR-ABL1 lymphoblastic leukaemia is characterized by the deletion of Ikaros. Nature 453, 110–114 (2008)

Multispecific Interactions
in Enzymatic Signalling Cascades

Daniel D. Seaton and J. Krishnan

Department of Chemical Engineering,
Centre for Process Systems Engineering and Institute for Systems
and Synthetic Biology, Imperial College London, London,
SW7 2AZ, UK
j.krishnan@imperial.ac.uk

Abstract. The reversible postranslational modification of proteins is a
ubiquitous feature of cellular signal transduction networks. In these systems, signalling is typically seen as resulting from the interaction between
an active enzyme and a downstream unmodified substrate. However, it
is known that in some cases the inactive form of an enzyme is also capable of binding the unmodified substrate, and that in other cases the
active enzyme is capable of binding modified substrate. In this paper,
we analyse the behaviour of a two-stage enzymatic cascade in which
these additional protein-protein interactions are possible. Without the
additional interactions, the model produces the standard ultrasensitive
switch-like behaviour. We find that inactive enzyme binding to unmodified substrate increases the ultrasensitivity of this switch, while active
enzyme binding to modified substrate results in the switch becoming
biphasic. These results indicate how important the rules governing the
occurrence of protein-protein interactions can be in determining the signalling behaviour of a pathway, even when particular protein-protein
interactions have no clear functional role.

Keywords: kinase signalling, ultrasensitivity, biphasic.

1 Introduction

Postranslational modifications are ubiquitous in cellular signal transduction networks. These are frequently viewed as systems in which activated enzymes provide a signal to downstream substrates through their enzymatic activity in a
unidirectional manner. There has recently been interest in extending this view
by considering the effects of enzyme uptake on such signalling cascades [18,15,9].
Here, we look to build on this work by considering a simple two-stage enzymatic
cascade where multiple additional protein-protein interactions are possible (see
figure 1).

The most basic model of signalling in this system involves just the interaction
of the active enzyme with the unmodified substrate. This is in keeping with a
model where activation of the enzyme's catalytic site is coupled to the enzyme's

M.A. Lones et al. (Eds.): IPCAT 2012, LNCS 7223, pp. 67–73, 2012.

ability to bind its substrates, as seen for example in the case of the kinase Src [5]. However, there are several additional interactions possible. First, the active enzyme may bind to the modified substrate. This corresponds to product inhibition, as seen for example in the case of $Ca{+}{+}$/calmodulin dependent protein kinases [7]. Second, the inactive enzyme may be able to bind to the unmodified substrate without catalyzing any reaction, and thereby interfering with the operation of the active enzyme. This is the case, for example, with the oxidative modification of a cysteine residue on protein tyrosine phosphatases such as MAP kinase phosphatase 3 (MKP3)[6,16]. By including these additional interactions in our model (as well, for completeness, the interaction between inactive enzyme and modified substrate), we are able to consider how they modify the basic behaviour of the cascade - the main objective of this paper.

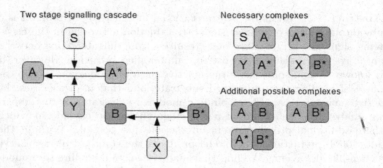

Fig. 1. The basic signalling system under consideration. Modified/activated species are indicated by an asterix. Solid lines indicate enzymatic conversion of species, while dashed lines indicate enzymatic regulation of a reaction. The necessary complexes are intermediates in the enzymatic reactions, while the additional possible complexes are not.

2   Results

In order to investigate the consequences of the additional interactions on signalling in this model, we begin by taking a basal set of parameters without any additional interactions. We then introduce various strengths of the additional interactions and observe how the signalling properties of the pathway are modified. We use one basal set of parameters, with which the model reproduces the the well known ultrasensitive response to the upstream signal (see figure 2). The details of the model and parameters used are available in the appendix. We will now consider the cases of product inhibition and inactive enzyme interference in turn.

2.1   Product Inhibition

In the case of product inhibition, we observe that at steady state there is a biphasic response to the signal - the output initially increases with the input signal, before decreasing to a steady value above zero (see figure 2). This is a clear qualitative change in behaviour compared to the case without product inhibition. Biphasic behaviour is typically observed in systems containing incoherent feedforward loops [8]. However, in this case there are no explicitly competing pathways. Instead, the biphasic behaviour is the result of a change in the primary role of the active enzyme, A*. At low signal levels, A* is primarily involved in modifying B. It does this without inhibition by B*, since both A* and B* are present at low levels. As the signal levels increase, however, A* begins to uptake significant quantities of B*. Under these conditions, A* is primarily involved in reducing the free B* available to act as output. Of course, if the complex A*B* is able to contribute to the output, this effect is lost (results not shown). This is key to the difference between the observations made here and those made in [13], in which outputs are taken in terms of total concentrations. While this may be the relevant variable in some situations, we argue that the free concentration is the most natural output variable. We further make a distinction between our use of "biphasic" - meaning a response that is increasing, then decreasing (as in [8]) - with that used in [13], in which "biphasic" is taken to mean a double-threshold behaviour.

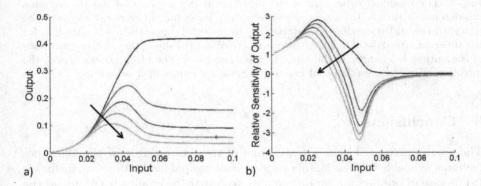

Fig. 2. The modified behaviour of the pathway in the presence of product inhibition is shown. (a) shows the steady-state input-output response curve of the concentration of free modified substrate, B*, to the total concentration of signal, S, for five different strengths of product inhibition. (b) shows the relative sensitivity of the output to changes in the input in the same cases. The black arrows denote the direction of increasing affinity of active enzyme for modified substrate.

2.2 Inactive Enzyme Interference

At steady state, the qualitative ultrasensitive behaviour of the pathway is not changed by the introduction of inactive enzyme interference. However, there is a clear change in the quantitative characteristics of the ultrasensitivity. The maximum relative sensitivity of the output to changes in input is larger, and occurs at higher input levels in the presence of inactive kinase interference (see figure 3). This is a result of the molecular titration of the unmodified substrate, B, by the inactive enzyme, A. This effect has been noted previously in other systems, notably multiphosphorylation switches [11], and genetic circuits [3,2].

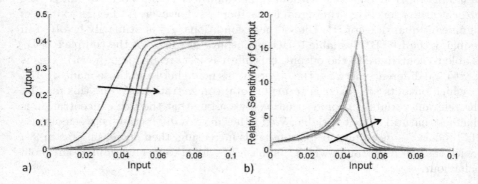

a)   b)

Fig. 3. The modified behaviour of the pathway in the presence of inactive enzyme interference is shown. (a) shows the steady-state input-output response curve of the concentration of free modified substrate, B^*, to the total concentration of signal, S, for five different strengths of inactive enzyme interference. (b) shows the relative sensitivity of the output to changes in the input in the same cases. The black arrows denote the direction of increasing affinity of inactive enzyme for unmodified substrate.

3   Conclusions

The results presented demonstrate some of the interesting new signalling behaviours available to these simple systems if additional interactions are included. In the case of product inhibition, we have seen that the qualitative nature of the behaviour can change dramatically from a monotonic response to a biphasic response. In the case of inactive enzyme interference, we have seen that the ultrasensitivity is enhanced as a result of the additional interaction. A key point to be made is that the models considered here are not very much more complex than what has been considered as the default model - they have a similar number of parameters, and the basic intuition about how the molecular interactions are occurring is the same. These results cover only the immediate consequences of two elaborations of the basic model for a particular set of parameters at steady state. Future work will be concentrated on two areas. In the first place, a more complete understanding of the model's behaviour is required. In the second place, an understanding of how other additional interactions may affect signalling is required.

A more complete understanding of the model's behaviour will come through systematic simulation and sensitivity analyses. For example, it will be necessary to investigate how additional interactions affect the signalling dynamics.

Furthermore, it will be necessary to investigate how the signalling properties of these pathways are modified by changes in parameters and concentrations of key components. There has been recent work into how a synthetic MAPK cascade's signalling properties vary with the concentrations of pathway components [12], and the robustness of signalling properties to perturbations in the concentrations of key components has been a subject of recent interest in a variety of other systems [1,17,10]. Clearly, the sensitivity and robustness of signalling properties to changes in such changes will be modified by the presence of additional interactions.

With regards to the effects of other interactions, we note that the model described here could be further expanded by including numerous other additional interactions. For example, it may be that a tertiary complex forms between S, A*, and B. This would simply mean that the activation of A by S is not simultaneous with dissociation of A from S, and that the presence of S doesn't prevent A* from binding to B. There are other similar examples possible, especially if the enzymes X and Y are included. Many of these cases may result in clear qualitative changes in the model behaviour, and it is an open question whether these can be related to one another and understood within a common framework.

Seen together, the effects on signalling of this set of interactions are interesting in that they operate purely through uptake of the species involved. While the effects of these particular interactions have not previously been investigated systematically, there are some interesting parallels with recent literature. The bidirectional signalling which results from uptake of a downstream component has been the subject of theoretical [18,15] and experimental investigation [9]. In addition, Ciliberto et al have shown that the binding activity of an inactive enzyme in positive feedback system can produce bistability [4], while Xing and Chen showed that binding of two substrates to the active enzyme allowed ultrasensitivity to occur in a simple push-pull network for which it couldn't otherwise occur [19]. This is in addition to a substantial body of work investigating the effects of enzyme uptake in multiphosphorylation switches [13,11,14]. This work is therefore an attempt to link these themes to the modelling of a well known signalling mechanism, with a view to building a more systematic understanding of these issues.

References

1. Acar, M., Pando, B.F., Arnold, F.H., Elowitz, M.B., van Oudenaarden, A.: A general mechanism for network-dosage compensation in gene circuits. Science 329(5999), 1656–1660 (2010), http://dx.doi.org/10.1126/science.1190544
2. Buchler, N.E., Cross, F.R.: Protein sequestration generates a flexible ultrasensitive response in a genetic network. Mol. Syst. Biol. 5, 272 (2009), http://dx.doi.org/10.1038/msb.2009.30
3. Buchler, N.E., Louis, M.: Molecular titration and ultrasensitivity in regulatory networks. J. Mol. Biol. 384(5), 1106–1119 (2008), http://dx.doi.org/10.1016/j.jmb.2008.09.079

4. Ciliberto, A., Capuani, F., Tyson, J.J.: Modeling networks of coupled enzymatic reactions using the total quasi-steady state approximation. PLoS Comput. Biol. 3(3), e45 (2007),
   http://dx.doi.org/10.1371/journal.pcbi.0030045
5. Cole, P.A., Shen, K., Qiao, Y., Wang, D.: Protein tyrosine kinases src and csk: a tail's tale. Curr. Opin. Chem. Biol. 7(5), 580–585 (2003)
6. den Hertog, J., Groen, A., van der Wijk, T.: Redox regulation of protein-tyrosine phosphatases. Arch. Biochem. Biophys. 434(1), 11–15 (2005),
   http://dx.doi.org/10.1016/j.abb.2004.05.024
7. Huynh, Q.K., Pagratis, N.: Kinetic mechanisms of ca++/calmodulin dependent protein kinases. Arch. Biochem. Biophys. 506(2), 130–136 (2011),
   http://dx.doi.org/10.1016/j.abb.2010.11.008
8. Kim, D., Kwon, Y.K., Cho, K.H.: The biphasic behavior of incoherent feed-forward loops in biomolecular regulatory networks. Bioessays 30(11-12), 1204–1211 (2008),
   http://dx.doi.org/10.1002/bies.20839
9. Kim, Y., Coppey, M., Grossman, R., Ajuria, L., Jimnez, G., Paroush, Z., Shvartsman, S.Y.: Mapk substrate competition integrates patterning signals in the drosophila embryo. Curr. Biol. 20(5), 446–451 (2010),
   http://dx.doi.org/10.1016/j.cub.2010.01.019
10. Krantz, M., Ahmadpour, D., Ottosson, L.G., Warringer, J., Waltermann, C., Nordlander, B., Klipp, E., Blomberg, A., Hohmann, S., Kitano, H.: Robustness and fragility in the yeast high osmolarity glycerol (hog) signal-transduction pathway. Mol. Syst. Biol. 5, 281 (2009), http://dx.doi.org/10.1038/msb.2009.36
11. Legewie, S., Schoeberl, B., Blthgen, N., Herzel, H.: Competing docking interactions can bring about bistability in the mapk cascade. Biophys. J. 93(7), 2279–2288 (2007), http://dx.doi.org/10.1529/biophysj.107.109132
12. O'Shaughnessy, E.C., Palani, S., Collins, J.J., Sarkar, C.A.: Tunable signal processing in synthetic map kinase cascades. Cell 144(1), 119–131 (2011),
   http://dx.doi.org/10.1016/j.cell.2010.12.014
13. Salazar, C., Hfer, T.: Kinetic models of phosphorylation cycles: a systematic approach using the rapid-equilibrium approximation for protein-protein interactions. Biosystems 83(2-3), 195–206 (2006),
   http://dx.doi.org/10.1016/j.biosystems.2005.05.015
14. Salazar, C., Hfer, T.: Multisite protein phosphorylation–from molecular mechanisms to kinetic models. FEBS J. 276(12), 3177–3198 (2009),
   http://dx.doi.org/10.1111/j.1742-4658.2009.07027.x
15. Seaton, D.D., Krishnan, J.: The coupling of pathways and processes through shared components. BMC Syst. Biol. 5, 103 (2011),
   http://dx.doi.org/10.1186/1752-0509-5-103
16. Seth, D., Rudolph, J.: Redox regulation of map kinase phosphatase 3. Biochemistry 45(28), 8476–8487 (2006), http://dx.doi.org/10.1021/bi060157p
17. Shinar, G., Milo, R., Martnez, M.R., Alon, U.: Input output robustness in simple bacterial signaling systems. Proc. Natl. Acad. Sci. U S A 104(50), 19931–19935 (2007),
   http://dx.doi.org/10.1073/pnas.0706792104
18. Vecchio, D.D., Ninfa, A.J., Sontag, E.D.: Modular cell biology: retroactivity and insulation. Mol. Syst. Biol. 4, 161 (2008),
   http://dx.doi.org/10.1038/msb4100204
19. Xing, J., Chen, J.: The goldbeter-koshland switch in the first-order region and its response to dynamic disorder. PLoS One 3(5), e2140 (2008),
   http://dx.doi.org/10.1371/journal.pone.0002140

Appendix

The model is described by the set of ordinary differential equations given below. This model includes all possible binary interactions between A and B, as shown in figure 1, although we only analyse the cases where the strengths of product inhibition and inactive enzyme interference are modulated. All concentrations and parameters are dimensionless. The initial conditions determine the quantities of S, A, B, X, and Y available during the simulations. The initial concentration of S was varied over the ranges shown in the figures, while all other initial concentrations were kept constant. In particular, $[A_0] = 3$, $[B_0] = 1$, $[X_0] = 0.05$, $[Y_0] = 0.1$. The parameters used were as follows: $k_{1f} = 10$, $k_{3f} = 10$, $k_{5f} = 50$, $k_{7f} = 10$, $k_{9f} = 0$, $k_{10f} = 0$, $k_{11f} = 0$, $k_{1r} = 1$, $k_{3r} = 1$, $k_{5r} = 1$, $k_{7r} = 1$, $k_{9f} = 1$, $k_{10f} = 1$, $k_{11f} = 1$, $k_2 = 0.1$, $k_4 = 0.1$, $k_6 = 0.5$, and $k_8 = 0.1$. In the case of product inhibition and inactive enzyme interference, the parameters k_{9f} and k_{10f}, respectively, took the values 0, 1, 2.5, 5, and 10. The initial conditions and parameters chosen for the basal case were chosen so that the model would reproduce the typical behaviour of an enzymatic cascade in a regime where enzyme uptake was significant.

$$d[S]/dt = -(k_{1f}[S][A] - k_{1r}[SA]) + k_2[SA]$$
$$d[A]/dt = -(k_{1f}[S][A] - k_{1r}[SA]) + k_4[XA^*] - (k_{10f}[A][B] - k_{10r}[AB])$$
$$-(k_{11f}[A][B^*] - k_{11r}[AB^*])$$
$$d[A^*]/dt = k_2[SA] - (k_{3f}[X][A^*] - k_{3r}[XA^*]) - (k_{7f}[A^*][B] - k_{7r}[A^*B])$$
$$+k_8[A^*B] - (k_{9f}[A^*][B^*] - k_{9r}[A^*][B^*])$$
$$d[B]/dt = k_6[YB^*] - (k_{7f}[A^*][B] - k_{7r}[A^*B]) - (k_{10f}[A][B] - k_{10r}[AB])$$
$$d[B^*]/dt = k_8[A^*B] - (k_{5f}[Y][B^*] - k_{5r}[YB^*]) - (k_{9f}[A^*][B^*] - k_{9r}[A^*][B^*])$$
$$-(k_{11f}[A][B^*] - k_{11r}[AB^*])$$
$$d[X]/dt = -(k_{3f}[X][A^*] - k_{3r}[XA^*]) + k_4[XA^*]$$
$$d[Y]/dt = -(k_{5f}[Y][B^*] - k_{5r}[YB^*]) + k_6[YB^*]$$
$$d[SA]/dt = (k_{1f}[S][A] - k_{1r}[SA]) - k_2[SA]$$
$$d[XA^*]/dt = (k_{3f}[X][A^*] - k_{3r}[XA^*]) - k_4[XA^*]$$
$$d[YB^*]/dt = (k_{5f}[Y][B^*] - k_{5r}[YB^*]) - k_6[YB^*]$$
$$d[A^*B]/dt = (k_{7f}[A^*][B] - k_{7r}[A^*B]) - k_8[A^*B]$$
$$d[A^*B^*]/dt = (k_{9f}[A^*][B^*] - k_{9r}[A^*][B^*])$$
$$d[AB]/dt = (k_{10f}[A][B] - k_{10r}[AB])$$
$$d[AB^*]/dt = (k_{11f}[A][B^*] - k_{11r}[AB^*])$$

$$(1)$$

Role of Physico-chemical Properties of Amino Acids in Protein's Structural Organization: A Network Perspective

Dhriti Sengupta and Sudip Kundu

Department of Biophysics, Molecular Biology and Bioinformatics,
University of Calcutta,
92 APC Road, Kolkata 700009, India
dhritisen@gmail.com, skbmbg@caluniv.ac.in
www.caluniv.ac.in

Abstract. The three-dimensional structure of a protein can be described as a graph where nodes represent residues and interactions between them are edges. We have constructed protein contact networks at different length-scales for different interaction strength cutoffs. The largest connected component of short-range networks exhibit a highly cooperative transition, while long- and all-range networks (more similar to each other), have less cooperativity. The hydrophobic subnetworks in all- and long-range networks have similar phase transition behaviours while hydrophilic and charged networks don't. Hydrophobic subclusters in long- and all-range networks exhibit higher occurrence of assortativity and hence higher communication ability in transmitting information within a protein. The highly cliquish hydrophobic nodes in long- and short-range networks play a significant role in bridging and stabilizing distantly placed residues during protein folding. We have also observed a significant dominance of charged residues cliques in short-range networks.

Keywords: Protein contact network, Largest cluster transition, Assortativity, Clustering coefficient.

1 Introduction

The interactome of amino acids within a protein's 3D structure can be represented as a network where amino acids are the nodes and interactions among them are considered as edges. This powerful network representation (called Protein contact networkor PCN) helps researchers to understand different aspects of a protein's structural organization and stabilization. Some of them are identifying key residues stabilizing protein's 3D structure, folding nucleus, mixing behavior of the amino acids etc [1–10]. Long-range interactions (interactions of amino acids placed at long distance in primary structure) within a protein play a distinct role in determining the tertiary structure of a protein, while short-range interactions (interactions of amino acids closely place in primary structure) largely contribute to the secondary structure formations. Researchers

M.A. Lones et al. (Eds.): IPCAT 2012, LNCS 7223, pp. 74–81, 2012.

have also studied the role of protein contact networks at different length scales of primary structure in protein folding and stability [11–13].

Vishveshwara *et al* have studied the change of largest cluster size with interaction strength cutoff [2] (that has been used to give a link among amino acids and hence to construct the network). They have observed the phase transition behaviour of the largest cluster size. Comparing the protein contact network with random model, they have concluded that the bond percolation resembles with random model (the probability of connection between two amino acids depends only on a specific I_{min}); however clique percolation cannot be achieved by random like behaviour [14, 15]. They have not addressed whether these phase transition behaviour of largest cluster size depends on the relative position of interacting amino acids in the primary structure.

Here, we have studied the change of largest cluster sizes; the mixing behaviour of nodes; overall cliquishness as well as preference of specific types of cliques over others in different subnetworks. We observe that the phase transition behaviours of long-range networks and short-range networks are different and the former have higher similarity with all-range networks. While the mixing behaviour of amino acids within all-range contact network is reflected in their long- and short-range networks, the hydrophobic subnetworks have a significant contribution in determining the overall mixing property of long-range networks. We also demonostrate the higher occurrence of hydrophobic residues' cliques in all- and long-range networks. On the other hand, cliques of charged residues are over-represented in short-range networks.

2 Methods

Dataset consists of large number of non-redundant protein structures from the Protein Data Bank. Each protein can be represented as a graph where its nodes (amino acids) are connected by edges based on the strength of non-covalent interactions between two amino acids which come within a distance of 5A.

Strength of interaction between two amino acid side chains is evaluated as a percentage given in [2] by:

$$INT(R_i, R_j) = \frac{n(R_i, R_j)}{\sqrt{N(R_i) \times N(R_j)}} \times 100 \qquad (1)$$

where, $n(R_i, R_j)$ is the number of distinct interacting pairs of side-chain atom between the residues R_i and R_j, which come within a distance of 5 $\mathring{A}$ (the higher cutoff for attractive Londonvan der Waals forces [16]) in the 3D space. $N(R_i)$ and $N(R_j)$ are the normalization factors for the residue types R_i and R_j. The network topology of such protein structure graphs depends on the cutoff (I_{min}) of the interaction strength between amino acid residues used in the graph construction. Any pair of amino acid residue (R_i and R_j) with an interaction strength of I_{ij}, are connected by an edge if $I_{ij} > I_{min}$. We varied I_{min} from 0% to 10%, and protein contact networks are constructed at these different cutoffs.

The PCNs generated by this method are all-range networks (ARNs). When two amino acids in such networks are separated by at least 10 residues in primary structure, it would be a part of long-range network (LRN), otherwise it would be part of short-range network (SRN) [3, 10]. We have also generated hydrophobic networks (BN), where only hydrophobic residues are the nodes and the links among themselves are considered as edges. Hydrophilic (IN) and charged networks (CN) are also constructed similarly [10]. Each of the networks is represented as an adjacency matrix. Any element of the adjacency matrix (A), connecting the ith and j^{th} nodes, is given as: $a_{ij} = 1$, if $i \neq j$ and nodes i and j are connected by an edge, the value is 0 if $i \neq j$ and nodes i and j are not connected or if $i = j$.

To study the tendency for nodes in networks to be connected to other nodes that are like (or unlike) them, we have calculated the Pearson correlation coefficient (r) of the degrees at either ends of an edge. Its value has been calculated using the expression suggested by Newman [17] and is given as

$$r = \frac{M^{-1} \sum_i j_i k_i - [M^{-1} \sum_i 0.5(j_i + k_i)]^2}{M^{-1} \sum_i 0.5(j_i^2 + k_i^2) - [M^{-1} \sum_i 0.5(j_i + k_i)]^2} \tag{2}$$

Here j_i and k_i are the degrees of the vertices at the ends of the ith edge, with $i = 1, ...M$. The networks having positive r values are assortative in nature.

The clustering coefficient (C) is a measure of local cohesiveness. (C_i) of a node i is the ratio between the total number of links actually connecting its nearest neighbors and the total number possible links between the nearest neighbors of node i. In other words,(C_i) enumerates the number of loops of length three maintained by a node i and its interconnected neighbors. It is given by

$$C_i = \frac{2e_i}{k_i(k_i - 1)} \tag{3}$$

Here e_i is the total number of edges actually connecting the ith node's nearest neighbors and k_i is the number of neighboring nodes of node i.

3   Results and Discussion

Here, the hydrophobic (BN), hydrophilic (IN), charged (CN) and all residues(AN) networks at three different length scales [long-range interaction networks (LRNs), short-range interaction networks (SRNs) and all-range interaction networks (ARNs)] for each protein have been constructed and analyzed at different interaction strength (I_{min}) cutoffs.

3.1   Transitions of Largest Clusters Sizes Depend on Length Scale and Physico-chemical Nature of Amino Acid Nodes

The Largest Connected Component (LCC) provides information on the nature and connectivity of the network. Researchers have studied LCC as a function of

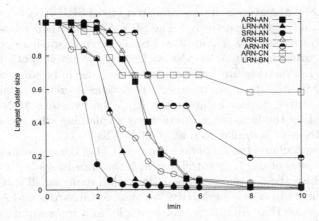

Fig. 1. Transition profile of different subnetworks. The size of largest connected component is plotted as a function of I_{min} for different subnetworks in a representative protein (PDB code: 1A2O). The subnetworks are - All-range all residue network (ARN-AN), Long-range all residue network (LRN-AN), Short-range all residue network (SRN-AN), All-range hydrophobic residue network (ARN-BN), All-range hydrophilic residue network (ARN-IN), All-range charged residue network (ARN-CN) and Long-range hydrophobic residue network (LRN-BN).

interaction strength ($I_{min}\%$), and analyzed phase transition behaviour of LCC from percolation point of view [2]. The critical interaction strength values below which the PCNs are almost completely connected and above which the PCNs split up into smaller subclusters, are also reported for a large number of globular proteins [2].

We have studied how the phase transition of LCC is affected by the physico-chemical properties of the nodes (BN, IN, CN and AN) and also whether the nature of transition of different subclusters depends on length scales (LRN, SRN and ARN). The normalized size of LCC as a function of I_{min} for different subnetworks are plotted in Figure 1. The particular I_{min} at which the cluster separates to form many small disjoint clusters or vice versa (i.e. the mid point of transition) is termed as $I_{critical}$.

While the LCC has been plotted as a function of I_{min} cutoff, we observe that the nature of transition in ARN-AN is closer to LRN-AN than SRN-AN. At I_{min} =0%, the largest cluster of ARN-AN is fully connected, i.e. it includes all the amino acids residues of a protein. The transition of LCC occurs within a narrow range of I_{min} for most (approximately 90%) of proteins. The values of proteins $I_{critical}$ vary from I_{min} =3% to 4.5%. In SRN-ANs, transition begins at a low I_{min} cutoff. Nearly 86% proteins have an $I_{critical}$ between 1% to 1.5% making the transition very sharp. On the other hand, the transition of LRN-ANs is slower than SRN-ANs but fasrer than ARN-ANs. The $I_{critical}$ values vary from 1.5% to 3% in approximately 88% proteins.

The clusters in SRNs are highly connected at lower I_{min} cutoffs. Although, the average cluster size of LRN-ANs at 0% I_{min} is lower than SRN-ANs (almost same as ARN-ANs), the average interaction strength of SRN-ANs is lower than LRN-ANs (2.56 and 2.86, respectively). The SRN-ANs lose their connectivities very rapidly with increase in I_{min} cutoff and break down to smaller sub clusters causing a steep transition. In ARN-ANs at lower I_{min} cutoff, both the long- and short-range interactions are involved in forming the fully connected large cluster. The contribution from short-range interactions decreases steadily with increase in I_{min} cutoff. So finally, at higher I_{min}, the resulting ARN-ANs are mainly connected by the long-range interactions, explaining why the nature of transition in ARN-ANs is similar to that of LRN-ANs.

Next, we have calculated the cooperativit index (CI) of the transitions [18]. CI is defined as the ratio of the I_{min} cutoff at which the transitions begins and the I_{min} cutoff at which the clusters just break down into many small sub-blusters. Lower CI value suggests less cooperativity. We observe LRN-ANs and ARN-ANs have a lower CI value than SRN-ANs. For example, in a representative protein 1A2O, SRN-AN show the highest average CI value (0.55), which is approximately 1.5 times of CI values of LRNs (0.33) and ARNs (0.29).

We have also studied the transitions of LCCs in in the ARN-BNs, ARN-INs and ARN-CNs. The ARN-INs, ARN-CNs and SRN-BNs do not show any sigle state transition. On the other hand, the LCC of ARN-BN and LRN-BN undergo single state transition and the behaviour is more inclined towrds that of ARN-ANs.

3.2 Assortative Mixing Behavior and Clustering Coefficients of Hydrophobic Residues in LRNs

To understand the mixing behavior of nodes, we calculated Pearson correlation coefficient values of different subnetwoks. The nodes with high degrees have tendencies to be connected with nodes having higher and lower degree in an assortative (positive) and disassortative (negative) networks, respectively. It is known that assortative networks percolate easily, i.e. information can be easily transferred through the assortative network as compared to a disassortative network [17]. We observed that ARN-ANs, LRN-ANS, SRN-ANs and most of the ARN-BNs are all assortative clusters (Table 1). In LRNs, assortativity is mainly contributed by hydrophobic residues (LRN-BNs have 90% subclusters with positive values while LRN-INs have 40% subclusters with negative values).

Thus, the mixing behavior of the amino acids is mainly influenced by the hydrophobic residues. One can expect that for any perturbation at the residue level, the necessary communication to the distantly located site would pass easily through the chain of hydrophobic residues. In addition, it indicates the hydrophobic residues' major involvement in the folding process of a protein. Protein folding is a cooperative phenomenon where communication amongst amino acids is essential, so that appropriate non-covalent interactions can form the stable native state structure [19]. It is also reported that the assortativities in ARNs and LRNs positively correlate to the rate of folding [20]. Here we show that the hydrophobic subclusters have the highest assortative mixing behavior in LRN and ARNs; and thus may indirectly indicate the hydrophobic residues'

Table 1. Average values of Pearson correlation coefficient ($\langle r \rangle$) and clustering coefficients ($\langle C \rangle$) of hydrophobic (BN), hydrophilic (IN), charged (CN), and all-amino-acids (AN) networks at different length scales viz. the long-range (LRN), short-range (SRN) and all-range (ARN) interaction networks are listed for $I_{min} = 0$

Length scale	Type	$\langle r \rangle$	$\langle C \rangle$
LRN	BN	0.130 ± 0.104	0.24 ± 0.05
	IN	-0.036 ± 0.186	0.14 ± 0.07
	AN	0.169 ± 0.067	0.17 ± 0.03
SRN	BN	-0.011 ± 0.171	0.3 ± 0.09
	AN	0.214 ± 0.058	0.36 ± 0.04
ARN	BN	0.27 ± 0.076	0.39 ± 0.04
	IN	0.151 ± 0.146	0.3 ± 0.07
	CN	0.144 ± 0.163	0.28 ± 0.08
	AN	0.304 ± 0.042	0.36 ± 0.02

major involvement in communicating necessary information across the network in the folding process of a protein and help in generating the topology of tertiary structure of a protein.

Clustering coefficient is the measure of cliquishness of a network. It is negatively related to folding rate [20], so that necessary local and global organization can be achieved through connectivities of the amino acid residues. The average clustering coefficient values are summerized in Table 1. Higher values of in BNs imply higher influence of hydrophobic residues in the folding rate of a protein. It is also evident from higher values of in LRN-BNs and ARN-BNs that hydrophobic residues interact in a more connected fashion, stitching different secondary, super-secondary structures and stabilizing the protein structure at the global level.

3.3   Clique Occurrences and Perimeters Covered by Cliques

The clustering coefficient enumerates number of loops of length three. These loops (or cliques) can be helpful in understanding internal structural organization of proteins. We have calculated the occurrences of all cliques and normalized them with respect to the number of hydrophobic (B), hydrophilic (I) and charged (C) residues present in the ARN and LRN-ANs, at different I_{min} cutoffs. The observations are given below. In LRN-ANs and ARN-ANs, about 97.7% of proteins show highest number of BBB cliques. With increase in interaction strength, fraction of charged residues cliques increases when compared to the other possible clique combinations.

Each clique of degree, k=3 is equivalent to a triangle where amino acids represent the vertices. Considering the positions of amino acids (vertices) in primary structure, we have calculated the perimeter of the cliques. Higher perimeter of a clique suggests that amino acids of the clique are distantly placed in primary structure and have come closer in 3D space, thus playing an important role in fixing the tertiary structures. For each protein, we have calculated the average values of the perimeters for each type of combination of the cliques in ARN-ANs, and identified the combination which covers maximum perimeter values. The results are summarized below: BBB cliques cover maximum perimeters, followed by CCC cliques. Interestingly, at higher I_{min} cutoff, more number of proteins has maximum perimeter coverage by CCC cliques, followed by BBB cliques.

This study strongly indicates about the involvement of charged residues (in addition to hydrophobic residues) in shaping the overall topology of a protein by bringing the distantly placed amino acid residues along a polypeptide chain closer in the 3D space.

4   Conclusions

The network properties of a protein depend on the physico-chemical nature of its constituent amino acids and also on their relative positions in the primary structure. The phase transition behaviour of the largest connected component of long-range network and its hydrophobic subnetworks are more similar (than others) to that of all-range all amino acid contact network within a protein. In addition, the clustering coefficients and assortative mixing behaviour of hydrophobic subclusters suggest the predominant role in protein's structural organization, stability and folding.

References

1. Bolde, C., Kovalcs, I.A., Szalay, M.S., Palotai, R., Korcsmalros, T., Csermely, P.: Network analysis of protein dynamics. FEBS Lett. 581, 2776–2782 (2007)
2. Brinda, K.V., Vishveshwara, S.: A network representation of protein structures: implications to protein stability. Biophys. J. 89, 4159–4170 (2005)
3. Greene, L.H., Higman, V.A.: Uncovering network systems within protein structures. J. Mol. Biol. 334, 781–791 (2003)
4. Dokholyan, N.V., Li, L., Ding, F., Shakhnovich, I.: Topological determinants of protein folding. Proc. Natl. Acad. Sci. USA 99, 8637–8641 (2002)
5. del Sol, A., Fujihashi, H., Amoros, D., Nussinov, R.: Residues crucial for maintaining short paths in network communication mediate signaling in proteins. Mol. Syst. Biol. 2 (2006); 2006.0019
6. Amitai, G., Shemesh, A., Sitbon, E., Shklar, M., Netanely, D., Venger, I., Pietrokovski, S.: Network analysis of protein structures identifies functional residues. J. Mol. Biol. 344, 1135–1146 (2004)
7. Vendruscolo, M., Dokholyan, N.V., Paci, E., Karplus, M.: Small-world view of the amino acids that play a key role in protein folding. Phys. Rev. E 65, 06191 (2002)
8. Kundu, S.: Amino acids network within protein. Physica A 346, 104–109 (2005)
9. Aftabuddin, M., Kundu, S.: Weighted and unweighted network of amino acids within protein. Physica A 39, 895–904 (2006)
10. Aftabuddin, M., Kundu, S.: Hydrophobic, hydrophilic, and charged amino acid networks within protein. Biophys. J. 93, 225–231 (2007)

11. Gromiha, M.M., Selvara, S.: Influence of medium and long-range interactions in protein folding. Prep. Biochem. and Biotechnol. 29, 339–351 (1999)
12. Go, N., Taketomi, H.: Respective roles of short- and long-range interactions in protein folding. Proc. Natl. Acad. Sci. USA 75, 559–563 (1978)
13. Selvaraj, S., Gromiha, M.M.: Role of hydrophobic clusters and long-range contact networks in the folding of $(\alpha/\beta)_8$ barrel proteins. Biophys. J. 84(3), 1919–1925 (2003)
14. Brinda, K.V., Vishveshwara, S., Vishveshwara, S.: Random network behaviour of protein structures. Mol. BioSyst. 6, 391–398 (2010)
15. Deb, D., Vishveshwara, S., Vishveshwara, S.: Understanding protein structure from a percolation perspective. Biophysical Journal 97(6), 1787–1794 (2009)
16. Tinoco, I., Sauer, K., Wang, J.C.: Physical Chemistry: Principles and Application in Biological Sciences. Prentice-Hall, Englewood Cliffs (2001)
17. Newman, M.E.J.: Assortative mixing in networks. Phys. Rev. Lett. 89, 208701–208704 (2002)
18. Segel, I.H.: Biochemical Calculations. John Wiley and Sons, New York (1997)
19. Maity, H., Maity, M., Krishna, M.M.G., Mayne, L., Englander, S.W.: Protein folding: the stepwise assembly of foldon units. Proc. Natl. Acad. Sci. USA 102, 4741–4746 (2005)
20. Bagler, G., Sinha, S.: Assortative mixing in protein contact networks and protein folding kinetics. Bioinformatics 23, 1760–1767 (2007)

Tailored Strategies for the Analysis of Metabolomic Data

Kristen Feher[1], Kathrin Jürchott[1], and Joachim Selbig[1,2]

[1] Institute of Biochemistry and Biology, AG Bioinformatics, University of Potsdam,
Karl-Liebknecht-Str. 24-25, 14476 Potsdam, Germany
[2] Max-Planck Institute for Molecular Plant Physiology, Am Mühlenberg 1, 14476
Potsdam, Germany
{feher,selbig}@mpimp-golm.mpg.de

Abstract. Differences in tissues arising from a single organism are attributable, at least partially, to differing metabolic regimes. A highly topical instance of this is the Warburg effect in tumour development, whereby malignant tissue exhibits greatly altered metabolism compared to healthy tissue. To this end, we consider the emergent properties of two metabolomic datasets from a human glioma cell line (U87) and a human mesenchymal stem cell line (hMSC). Using a random matrix theory (RMT) approach, U87 is found to have a modular structure, whereas hMSC does not. The datasets are then compared using between groups comparison of principal components, and finally, a group of metabolites is found that remains highly correlated in both conditions.

1 Introduction

The study of biological systems from a metabolic perspective can be approached in a number of different ways. The information about which reactions each metabolite is involved in can be contained in a stoichiometric matrix S, which forms the basis for a variety of models based on the mass conservation law ([4]): $\frac{\partial \rho}{\partial t} = \mathbf{i} + S \cdot \nu - \mathbf{o}$, where ρ is the vector of metabolite concentrations, ν are the reaction fluxes, $\mathbf{i}$ is the input vector of fluxes and $\mathbf{o}$ is the output vector of fluxes. Under steady state conditions, i.e. $\partial \rho / \partial t = 0$, the solution space can be estimated yielding the whole set of stable metabolic fluxes that are possible, given a set of biochemical contraints S [11].

It can be theoretically shown that when considering fluctuations in a steady state solution of S, certain pairs of metabolites exhibit high correlation that can be considered a 'signature' for that steady state [1,13]. Furthermore, it is hard to find intuitive heuristic explanations for high correlations, as they rarely map to neighbours on the metabolic map. Instead, correlations can be interpreted as being caused by co-reponse to concentration of a common enzyme, chemical equilibrium, and mass or moiety conservation [1]. Changes in correlation between experiments can be attributed to changes in regulation [1]. In general, high correlations are not solely attributable to the reactions they participate in, but are emergent properties of the entire system. Both [1] and [13] consequently give

M.A. Lones et al. (Eds.): IPCAT 2012, LNCS 7223, pp. 82–89, 2012.

clear warnings against using coexpression of metabolites as a putative proxy for coregulation in the same manner as transcriptomic data. A metabolite correlation network can be constructed to decipher metabolomic data, however certain refinements can be made.

The individual metabolites are clearly not independent. Formally, an experiment is an observation of multivariate random variable $x = (x_1, x_2, \ldots, x_p)$, where x_i is the expression level of the ith metabolite for $i = 1, 2, \ldots, p$. In general the probability distribution function (PDF) of x is not the product of its marginal PDFs, i.e. $f(x) \neq f_1(x_1)f_2(x_2)\ldots f_p(x_p)$, where f, f_i is the PDF. We can say that $x_1, \ldots, x_p$ has a joint PDF (JPDF) f, emphasising that $x_1, \ldots, x_p$ have a complex dependence structure, and f is defined through the constraints of the metabolic network. The correlation matrix partially encodes the dependencies amongst the x_i, and this can be theoretically derived for a steady state with fluctuations via S [13]. Alternatively it can be estimated via the sample correlation matrix, derived from high throughput data.

A theoretical correlation matrix could be evaluated against a sample correlation matrix, for example to check whether the currently known S is correct, and this is where a correlation network might be used. However, given the complex emergent dependence structure, its richness might be missed by examining only significant pairwise correlations, as valuable information about the arrangement of 'moderate' correlation is lost [1]. Because the correlation matrix does not have a straightforward mapping back to the underlying network, a correlation network is simply an abstraction describing $f(x)$. Additionally, the location of $f(x)$ in $\mathbb{R}^p$ (equivalently the solution space of S) is also no longer explicitly considered.

Random matrix theory (RMT) is well suited to describing the emergent properties of systems with complex interactions amongst constituents via their eigenvalue spectrums [9]. It is also relevant to covariance estimation when $n < p$, where n is the number of observations or experiments, which is generally the case in high-throughput experiments (e.g. [5]).

In this paper, we consider the emergent properties of the U87 and hMSC datasets via their eigenvalue spectrums. The structure found here is also reflected in the conventional analysis of [6]. However, high correlations occur between metabolites with no obvious relationship, motivating a more holistic characterisation of the data, in preparation for integration with theoretical predictions. RMT methods are employed to choose a correlation network threshold of both hMSC and U87 in a principled way [3]. U87 is found to have a modular structure, whereas hMSC does not. The data-derived solution spaces for both hMSC and U87 are compared using between groups comparison of principal components [7], and results are assessed using an RMT style approach. Finally, a group of metabolites is found that remains highly correlated in both conditions.

2   Methods

17 samples each of human glioma cell line U87 and human mesenchymal stem cells were cultured, and metabolites were subsequently extracted and measured using gas chromotography mass spectrometry. Full details can be found in [6].

The structure of an $n \times p$ dataset X can be partially characterised via the eigen-value spacing distribution of thresholded sample correlation matrices, whereby RMT results state that a block diagonal matrix exhibit exponentially distributed spacing. The correlation matrices of each dataset are probed and block diagonal structure are found following the methods in [3].

Given two sets of observations of the same random variable: $n_1 \times p$ data matrix A and $n_2 \times p$ data matrix B, the difference between the subspace each matrix occupies can be partitioned through matrix W via between groups comparison of principal components. First define V_A and V_B to be the first r right singular vectors of A and B respectively. Then define matrix W such that [7]:

$$W = V_A V_B^T V_B V_A^T, \tag{1}$$

where W is an $r \times r$ matrix. Let $\mathbf{l}_i, i = 1, 2, \ldots, r$ be the eigenvectors of W with associated eigenvalues λ_i, then alternative bases for A and B can be respectively defined such that:

$$\mathbf{a}_i = V_A^T \mathbf{l}_i, \tag{2}$$

$$\mathbf{b}_i = V_B^T V_B \mathbf{a}_i, \tag{3}$$

for $i = 1, 2, \ldots, r$. Note that these bases *do not* correspond to directions of maximum variance, rather they correspond to a partitioning of difference between the two subspaces, whereby the angle $\cos^{-1}(\lambda_i)^{1/2}$ between $\mathbf{a}_i$ and $\mathbf{b}_i$ is minimised.

3   Results

In this section, the structure of the individual correlation matrices for both hMSC and U87 are probed, in order to gain insight into the dependence structure under both conditions. The hMSC dataset does not have a modular structure, as the eigenvalue spacing is never exponentially distributed. However, the U87 does exhibit signs of a modular structure at a correlation threshold of 0.75. There are 120 metabolites out of 139 in total with at least one coefficient greater than 0.75.

Four large clusters can be found in U87 (three small groups of size ≤ 3 are omitted from discussion here), corresponding to four main distinct signals which are not orthogonal, and hence are not obvious when PCA is performed over an entire dataset. The grouping of U87 is shown in a heatmap in Figure 1, along with hMSC, in which the same grouping has been induced. This characteristic of the U87 dataset can be interpreted that there are four groups of highly coordinated metabolites, yet those groups are independent from each other in the sense that they are highly distinct (and not that they are orthogonal).

Both datasets have comparable amounts of high correlation, but distinct types of arrangement. Overall, the U87 correlation network (but not necessarily the underlying metabolic network) could be qualitatively described as being 'coordinated' and having local 'connectivity', while the hMSC correlation network is not 'coordinated' and shows global 'connectivity'.

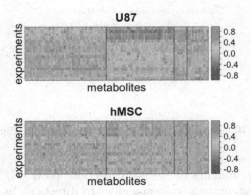

Fig. 1. Top: heatmap of four main clusters of U87. Bottom: heatmap of hMSC, with the induced U87 clustering. Note: it should be emphasised that the columns are centered and standardised to have unit length, hence red/green values indicate deviation from the column mean and all columns are on the same scale.

In the sense that U87 is 'coordinated' and locally 'connected', clusters are responding to stimuli as a single unit, due to the high mutual correlation occuring almost exclusively within clusters. Following [1], each cluster could putatively be explained by a large response to concentration of a common enzyme over all biological replicates, or by being in chemical equilibrium. Any two clusters which show an overall negative correlation between them could potentially be explained by a mass conservation relationship. It should be noted that the correlation matrix cannot identify the specific processes responsible, but rather interpretation of the correlation structure can be used to guide further analysis.

In contrast, hMSC is not 'coordinated' in the sense that the metabolites exhibit a much greater variety of responses that can't be distilled into a small number of signals, and shows global 'connectivity' because high correlation occurs widely across many different pairs of metabolites. One possible interpretation is that different pairs of metabolites co-respond to concentrations of different enzymes under different biological replicates, and so there is greater variability under fluctuation.

The theoretical JPDF $f(x)$ resulting from the constraints of S is unknown, however some of its properties can be inferred from the data. With only two datasets, it is not possible to make any statements about the entire solution space of S, but rather we compare the two subspaces of $\mathbb{R}^p$ defined by hMSC and U87. When multivariate data arises from such a constrained system as a metabolic network, it is not sensible to presume that the observations will be uniformly distributed over $\mathbb{R}^p$, but rather that they will be concentrated in certain regions corresponding to allowable solutions or states of the network.

Firstly, considering each dataset separately, their eigenvalue spectrums are plotted in Figure 2, represented as proportion of variance. Observations that are drawn from an identity covariance matrix are equally probable to occur in all

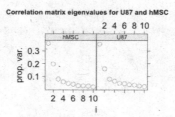

Fig. 2. First 10 eigenvalues of the correlation matrix hMSC and U87

directions of $\mathbb{R}^p$ [5]. Judging by visual inspection, both U87 and hMSC have at least 2 leading eigenvalues, corresponding to directions in which it is more probable to find an observation.

Next, it would be desirable to know whether the observations in each dataset originate from the same region of $\mathbb{R}^p$, by comparing the column spaces of each data matrix. Each observation is from the same multivariate random vector, i.e. each dataset observes the same set of variables, and both have the situation of $n < p$. Thus the column space of each will be of dimension n, constituting a subspace of $\mathbb{R}^p$. Following recommendations in [1], the datasets are standardised and treated separately as different states are likely to be on different scales; whereas if they were compared and clustered as one dataset for instance, spurious results are likely.

Define the U87 dataset as A and the hMSC dataset as B. Then new bases are found for A and B such that each successive direction accounts for an increasing amount of difference between A and B. The largest eigenvalue of W corresponds to the smallest possible angle between a vector constrained to lie in A and another vector constrained to lie in B. Using this technique, the angles corresponding to the eigenvalues of W are ($46.3°, 55.9°, 59.7°, 59.9°, 61.8°, 63.0°, 64.4°, 66.1°,$ $68.0°, 69.6°, 69.7°, 70.4°, 70.5°, 70.9°, 71.4°, 71.7°, 71.8°$).

The eigenvalues of W associated with A and B are assessed using a RMT style approach. The above technique was originally intended for the situation $n > p$. Each subspace would then be described by $k < p$ components, and k increased to understand how the two spaces overlapped. When $k = p$, this becomes the trivial case with exactly the same basis, and the k eigenvalues of W equal to 1, as the system in completely determined. In contrast, when $n < p$, the basis cannot be fixed when $k = p$ as there are not enough observations, and hence RMT can be used to understand the average behaviour of the eigenvalues of S. For comparison to the data, define a null hypothesis $H_0 : \Sigma_A = \Sigma_B = I$, i.e. both sets of observations are drawn from population identity covariance matrices. The results of a simulation study are displayed in Figure 3, where 1000 random pairs of matrices are generated, and the mean eigenvalues are plotted. H_0 has one large eigenvalue and the rest are small, corresponding to an angle of $22°$ and about $70°$. Under H_0, observations are equally likely in all directions, but because $n < p$, full 'coverage' of $\mathbb{R}^p$ can't be obtained, so once the first pair of axes are fixed, the rest have many degrees of freedom, which explains why the first angle is low

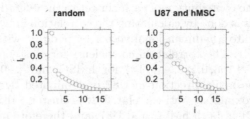

Fig. 3. Left: plots of the average eigenvalues of W under $H_0 : \Sigma_A = \Sigma_B = I$. Right: actual data eigenvalues from W.

and the rest are high. By comparison, the data's first angle is $44°$, indicating that A and B are drawn from separate regions of $\mathbb{R}^p$. Moreover, A and B are not uniformly distributed, as judged by their respective eigenvalue spectrums, indicating concentration in certain directions combined with separation. This is indicative of different combinations of correlated variables being active in both A and B. In addition, the other angles are lower than can be expected by H_0, indicating that there aren't as many degrees of freedom from which to draw observations, and this could be a direct consequence of the constraints of the underlying metabolic network.

A null hypothesis that observations are drawn from a population identity covariance matrix serves as an initial benchmark, but the eigenvalue spectrum of A and B shows is clearly different to that of the identity matrix. To check how well A and B can be discriminated, the row labels (experiments) are randomly permuted 1000 times, and the above test is repeated. Under permutation, 95% of the first angles are less than $40°$ and 99% are less than $42°$, while the data angle is $46°$. This indicates that A and B are able to be discriminated, or alternatively, the observations are drawn from different regions of $\mathbb{R}^p$. While this has shown that hMSC and U87 are able to be discriminated, this method delivers more information than discrimination, and this is explored further in the next section.

So far, it has been demonstrated that U87 has a strong modular structure, whereas hMSC doesn't. Between groups comparison of principal components has found that the datasets each arise from separate and distinct regions of $\mathbb{R}^p$. These results can be synthesised to understand which metabolites show similar correlation patterns under both conditions, as differential correlation can point to different regulatory regimes being active. Such an analysis is carried out in [10], but this was done by considering pairs of metabolites individually. The structure of $f(x)$ is the emergent result of interactions between all constituents in the system, and so it is more appropriate to consider the datasets as wholes.

The subspaces arising from each dataset are compared by relating its W basis to its PCA basis. Firstly, the correlation between the first W-basis vector $\mathbf{a}_1$ and the first principal component vector (PC1) for U87 is ~ 0.9656, indicating that a vector in the space defined by U87 is highly constrained to come from a single direction; while the correlation between the first W-basis vector $\mathbf{b}_1$ and PC1 for hMSC is ~ 0.5842.

Next, we check the contribution of each cluster to the overall principal component directions. This is done by calculating the correlation between each cluster's PC1 (transformed into a p-dimensional vector by padding with zeros) and each dataset's overall PC1. The correlation coefficients for clusters 1, 2, 3, 4 respectively are $\{0.635, -0.163, 0.122, -0.113\}$ for hMSC and $\{-0.847, 0.749, 0.121, 0.133\}$ for U87. The main direction in U87 is dominated by clusters 1 and 2, which are also approximately anti-correlated. It is cluster 1 that is most aligned with the overall PC1 in both hMSC and U87, and therefore makes the greatest contribution to the smallest separation between the 2 datasets. This common feature of both datasets indicates the possibility of a 'core' set of metabolites (at least under the two conditions considered) that always respond as a cohesive whole. Clusters 2-4 of hMSC aren't well correlated with PC1, and PC1 is only moderately correlated with b_1, suggesting there is much less constraint on the space of hMSC.

Interestingly, the two largest clusters identified in U87 have a tendency to include metabolites from different classes. Cluster 1 is dominated by lipids including fatty acids and cholesterol, while amino acids and TCA cycle intermediates can be found in cluster 2. Cluster 1 members remain highly correlated over hMSC and U87 glioma cells. This agrees with analysis performed on the known metabolites using the same data set in [6] where a substructure consisting of five lipid compounds is shared in hMSC and U87 glioma cells.

4   Conclusion

Much theoretical work has been done to characterise the emergent properties of metabolic networks. In particular, it can be shown that high correlation rarely occurs between neighbours on the metabolic map, and this must be kept in mind when interpreting metabolite correlation. Papers that address this [1,10,12] consider pairs of metabolites, but this has potential to obscure the emergent properties. In this paper, we have used RMT based methods to examine datasets as complete wholes via their eigenvalue spectrum. In the process, both their dependence structure and solution spaces have been characterised. We have found aspects that are consistent with a more biologically focussed analysis in [6], but also address the problem of non-obvious metabolite pairs with high correlation.

An advantage of applying RMT techniques to assess modularity is that it can guide the choice of analysis. For example, the lack of clear modularity structure in hMSC means that it is probably not meaningful to try clustering the data, but visual inspection of Figure 1 suggests it might be a good candidate for biclustering [2,8]. Subspaces defined by significant biclusters could then be further compared to U87 in a manner analogous to that outlined in this paper.

This is a two-way process. Not only can the theoretical predictions guide the analysis of high throughput data, but it is also important to evaluate theoretical predictions against the data. Before this can be done, a thorough understanding of the data's structure is needed, but this is not a trivial task in high dimensions with complex dependence structure. This paper can therefore be seen as a preliminary attempt to decompose data guided by theoretical predictions as a basis for future work. It is by no means complete.

Data analysis in the context of theoretical predictions can also stimulate further experimental design, by demonstrating the rich and detailed information that can be found from the careful dissection of a dataset. For instance, the methods outlined here could easily be extended to (many) more than two experimental steady state conditions. It would allow for a more complete enumeration of the combinatorics of the underlying metabolic map, i.e. the possible subsets of correlated metabolites under subsets of conditions. This could complement theoretical work in [13], by empirically creating a map between S and $f(x)$. In the context of cancer metabolomics, the fact that a steady state should have its own correlation 'signature' could eventually be used as a biomarker for diagnostic purposes. In short, theory and data analysis here are very synergistic, and should bring about new biological insight.

References

1. Camacho, D., de la Fuente, A., Mendes, P.: The origin of correlations in metabolomics data. Metabolomics 1, 53–63 (2005)
2. Cheng, Y., Church, G.: Biclustering of expression data. In: Proceedings of the Eighth International Conference on Intelligent Systems for Molecular Biology, pp. 93–103. AAAI Press (2000)
3. Feher, K., Whelan, J., Müller, S.: Assessing modularity using a random matrix theory approach. Stat. Appl. Genet. Mol. 10(1), 44 (2011)
4. Heinrich, R., Schuster, S.: The regulation of cellular systems. Chapman & Hall (1996)
5. Johnstone, I.M.: On the distribution of the largest eigenvalues in principal component analysis. Ann. Stat. 2, 295–327 (2001)
6. Jürchott, K., Guo, K.T., Catchpole, G., Feher, K., Willmitzer, L., Schichor, C., Selbig, J.: Comparison of metabolite profiles in U87 glioma cells and mesenchymal stem cells. Biosystems 105(2), 130–139 (2011)
7. Krzanowski, W.: Between-groups comparison of principal components. J. Am. Stat. Assoc. 74, 703–707 (1979)
8. Madeira, S.C., Oliveira, A.L.: Biclustering algorithms for biological data analysis: a survey. IEEE ACM T. Comput. Bi. 1, 24–45 (2004)
9. Mehta, M.L.: Random Matrices. Elsevier (2004)
10. Morgenthal, K., Weckwerth, W., Steuer, R.: Metabolomic networks in plants: transitions from pattern recognition to biological interpretation. Biosystems 83, 108–117 (2006)
11. Schellenberger, J., Palsson, B.O.: Use of randomized sampling for analysis of metabolic networks. J. Biol. Chem. 284, 5457–5461 (2009)
12. Steuer, R.: On the analysis and interpretation of correlations in metabolomic data. Brief. Bioinform. 7, 151–158 (2006)
13. Steuer, R., Kurths, J., Fiehn, O., Weckwerth, W.: Observing and interpreting correlations in metabolomic networks. Bioinformatics 19, 1019–1026 (2003)

Evolving Locomotion
for a Simulated 12-DOF Quadruped Robot

Gordon Klaus, Kyrre Glette, and Mats Høvin

University of Oslo, Norway
Department of Informatics
{gordonk,kyrrehg,matsh}@ifi.uio.no

Abstract. We demonstrate the power of evolutionary robotics (ER) by comparing to a more traditional approach its performance and cost on the task of simulated robot locomotion. A novel quadruped robot is presented, the legs of which – each having three non-coplanar degrees of freedom – are very maneuverable. Using a simplistic control architecture and a physics simulation of the robot, gaits are designed both by hand and using a highly parallel evolutionary algorithm (EA). It is found that the EA produces, in a small fraction of the time that takes to design by hand, gaits that travel at nearly twice the speed of the hand-designed one.

Keywords: evolutionary robotics, simulation, locomotion.

1 Introduction

Robot design can be complex. To design by hand, one often needs skills and experience with mechanics in general and with the particular robot and its dynamics. The space of possible solutions is enormous, so, to make the problem tractable, the designer is forced to make simplifying assumptions, for example by dividing the problem into more manageable pieces and limiting the dependencies among them; furthermore, he is typically biased towards modern engineering conventions and his individual experience. As a result, many potential solutions are not even considered. Thus, we cannot in general expect to design an especially good, let alone optimal, robot by traditional methods – perhaps not even with a very significant investment of effort.

Evolutionary robotics (ER) [1,2] solves this problem, to a great extent. When using ER, the necessary problem-specific competencies consist only of what is required to construct and evaluate the robot. One must still define, and thus limit, the solution space, but this can done in a manner unmotivated by the amenability of the resulting problem to manual analysis; bias can thus be freely controlled and nontraditional architectures explored. ER permits the designer to focus more of his energy on high level goals – i.e., on defining a fitness function – and it allows him to explore the vast space of possible solutions, including complex designs otherwise unlikely to be considered: for example, tensegrity structures and soft muscle-like actuators [3,4,5]. Whole families of robot architectures can be explored by allowing morphological parameters to vary within the optimization process [6,7,5].

M.A. Lones et al. (Eds.): IPCAT 2012, LNCS 7223, pp. 90–98, 2012.

While in its barest form ER involves evaluating candidate solutions in the real world, there is an enormous benefit to be gained by using a simulator. With sufficient computing power, many individuals can be evaluated in simulation in the time it would take to evaluate one in reality, while at the same time eliminating mechanical wear and unreliable human intervention. But there is no free lunch: in order to achieve this speedup, the simulator must sacrifice some accuracy; this yields a disparity between the simulated and real behaviors which has been termed the "reality gap" '[8]. Some efforts are currently underway to deal with this issue [9,10]. The other outstanding issue in ER (and with EAs in general) is the difficulty with scaling it to very complex designs while maintaining a reasonable time to convergence. Both of these issues are discussed in [1].

ER is far from perfect but it shows a great deal of promise. In this paper, we just begin to scratch the surface of its capabilities, laying a foundation for more in-depth investigations. We aim to demonstrate the effectiveness of ER by using it to design control systems for the simulated locomotion of a novel quadruped robot on flat ground.

In the next section we present the robot, the simulator, the EA, and the hand-designed gait. Thereafter, the results are discussed. Finally, we conclude the paper and describe our plans for future work.

2 Implementation

The robot that is the focus of this paper is a quadruped that has been designed in our lab and which is currently being constructed. As can be seen from the rendering in Figure 1, its body is an inverted rectangular pyramid and each of its legs consists of three linear actuators meeting at a spherical foot. The three non-coplanar degrees of freedom in each leg allow the foot to move freely in a large volume of space, providing the strength and wide range of motion that will be requisite for its ultimate role as a climbing robot; this should also make it an adept walker.

To simulate the robot's motions, the PhysX [11] physics simulation software library was used. A model of the robot, shown in Figure 2, was created in PhysX to capture the salient aspects of its design such as the types and connectivity of its joints and its rough shape. The ultimate goal is, of course, to make the simulation as realistic as possible so that designs can be transferred into the real robot, when it exists. Until that time comes, however, there is little immediate need to ensure that the simulation is identical to the expected reality.

A very simple mechanism was implemented to control the robot's motions. The target length of each actuator was set according to a periodic function of time of the form shown in Figure 3. Two parameters $\in [0,1)$ determine the exact shape of each control signal: the attack phase and the release phase, at which points the target length is set to its maximum and minimum values, respectively. Each actuator tries to drive towards its target length with a constant speed although opposing forces may slow it down. All of the actuator controllers operate at the same fixed frequency of 0.3 Hz but each actuator has its own phase

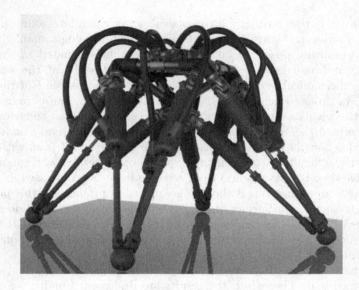

Fig. 1. A rendering of the robot model

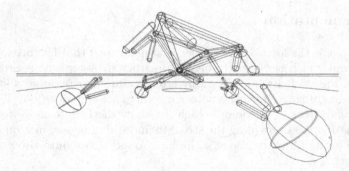

Fig. 2. The PhysX model of the robot. Only the shapes used for collision detection are shown, hence the gaps in the legs where the actuators (which have no shape - they are simply constraints) reside.

parameters; an entire controller is therefore defined by 24 phase values – two for each of the three actuators in each of the four legs. Such a simple controller was chosen for three reasons: (1) It is observed in nature that undisturbed gaits on flat ground are simply periodic; (2) to make it easier to hand-design gaits for comparison; and (3) to limit the size of the search space for the evolutionary algorithm. Also, simplicity is typically a very good place to start.

To evaluate a candidate controller, it was used to drive the robot for 16 seconds of simulated time; its fitness was calculated as its average speed over that interval, or zero if the robot fell over. Physics simulation is not particularly cheap – a single evaluation took on the order of one second to complete; so, rather than

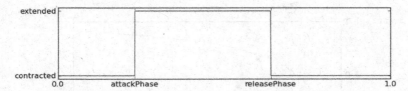

Fig. 3. The actuator control function

making it an exercise in patience, an effort was made to parallelize evaluations as much as possible. An incremental/steady state evolutionary algorithm (EA) was implemented such that each of the 150 CPUs in a small computing cluster was constantly under heavy load for the duration of a run of the algorithm; this system performed a very tolerable 60 evaluations per second. The controllers were encoded as a vector of 24 real numbers in the periodic interval $[0, 1)$. The initial population was generated randomly from a uniform distribution. New individuals were created by selecting from a population of 20 using fitness proportionate selection and mutating by adding a 24-vector of normally distributed values with mean zero and standard deviation 0.04. After being evaluated, the new individual was inserted into the population and the least fit (which could be the one just inserted) was removed. Ten thousand individuals were evaluated in a single run.

One gait was additionally hand designed for the purpose of comparison. This walker oriented its motion as a normal quadruped would with two legs to each side. In each leg, the actuators operated out of phase to produce a circle-like motion; legs on opposite sides circled in opposite directions. All legs moved out of phase to produce something resembling a typical walking gait.

3 Results and Discussion

A total of 100 evolutionary runs were performed. The average best fitness achieved was 1.33 m/s, with a standard deviation of 0.12 m/s. The best fitnesses of the populations during evolution, averaged over all 100 runs, are plotted in Figure 4. It appears that, after 10000 evaluations, the EA has not yet converged, suggesting that even better solutions might be found if more evaluations are performed.

The fastest evolved gait achieved an average speed of 1.55 m/s, more than twice that of the hand-designed gait, which moved at 0.68 m/s. Even the average evolved gait was nearly twice as fast as the hand-designed one. In the following paragraphs we discuss the behavior of the hand-designed gait, an evolved gait of average speed, and the fastest evolved gait.

It is immediately clear from the trajectory plots in Figure 5 that all of the gaits are oriented like a typical quadruped with two legs to each side rather than in some other manner such as with single leading leg, which is unsurprising given that nature has found the same solution. Also apparent is that all the

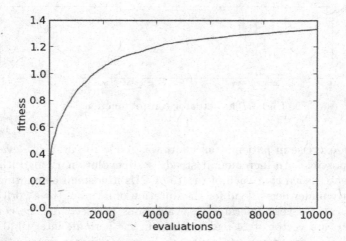

Fig. 4. The best fitnesses during evolution, averaged over 100 runs

paths have a steady curve to them, the evolved gaits most obviously. There is no bias toward symmetry in the controller representation, so this behavior is perfectly understandable; but even the hand-designed gait, which was designed with perfect symmetry, has a slight bend to it, suggesting that this robot is not perfectly balanced (in some sense of the word). Although it is somewhat difficult to see from these rather short trajectories, the fastest gait curves less than the average gait and, promisingly, to a degree comparable with the hand-designed gait (which is as straight as we might expect to achieve).

From the gait patterns in Figure 6 we can see a trend toward light-footedness – that is, the slowest gait has the most contact with the ground, dragging one of its feet nearly continuously; the average gait also drags a foot, but its other feet touch the ground for shorter periods; and the fastest gait steps quickly with all of its feet. Additionally, the faster gaits appear to effectively exploit the relatively low dynamic friction with the ground by sliding more often.

The more general character of the gaits can only be seen in their live action[1]. Most striking is the manner in which the evolved controllers cause the robots to lunge forward with their front legs rather than simply stepping; and how the timing and placement of the steps are done in such a manner that the robot's forward momentum is largely maintained.

Most relevant to the thesis of this paper is the amount of time and effort that went into producing each of the gaits. To design the one gait by hand took about 30 minutes, given no prior experience in designing gaits (or any robot controllers for that matter). It was a straightforward task to produce such a simple motion using the very simple control architecture provided (attack and release phases). However, it quickly became clear that to produce a more sophisticated controller,

---

[1] Videos can be found at http://folk.uio.no/gordonk/ipcat2012/.

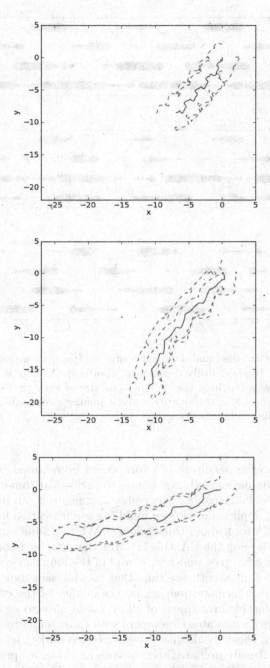

Fig. 5. Robot trajectories using the hand-designed controller (top), an average evolved controller (middle), and the best evolved controller (bottom). The solid line indicates the body position and the dotted lines indicate the feet.

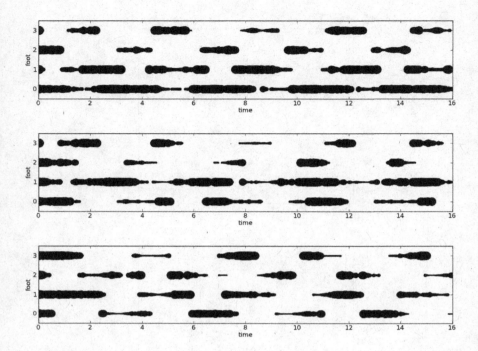

Fig. 6. Gait patterns for the hand-designed controller (top), an average evolved controller (middle), and the best evolved controller (bottom). A circle is drawn for each time step that a foot is touching the ground. The size of each circle corresponds to speed: a large circle indicates stationarity while a smaller circle indicates movement; therefore, a thinner line indicates dragging.

a significantly greater expenditure of effort would be required. An analogy to programming is illuminating: While it is easy to perform arithmetic calculations using a very low level language (e.g., assembly language), it can be prohibitively difficult to write an application of even middling sophistication using it.

Each run of the EA took about three minutes. While it additionally took quite a bit more time to develop the EA, this is a cost that can be amortized over the many uses of the system, past and future, and in the long run goes to zero.

Thus, the EA took one-tenth the time that hand-design took and yielded a gait twice as fast. Furthermore, judging by the quality of the evolved gaits, it seems that a considerable investment of effort would have to be made in order to manually construct a gait able to compete with those found by evolution; and the product of that effort would likely be somewhat task- and robot-specific, whereas the EA is already prepared to deal with an arbitrary problem.

This single simple result, although encouraging, should not lead one to greatly discount the value of traditional robot design in favor of evolutionary methods; each, of course, has its strengths. For example, even when an EA appears to have converged, we can never be sure that it has found the global optimum; but if we use the results of the EA to inform or inspire a more exact method then perhaps we can expect more. This approach is known as evolutionary-aided design.

4   Conclusion and Future Work

In this paper we have presented a novel robot design which should be suitable for the task of climbing and which is used here for the task of simulated locomotion. A parallel EA was created for rapidly evolving controllers for the robot, but which is in no way limited to this particular robot architecture or task. Gaits were designed both by hand and using the EA, and the advantages of evolution were apparent: It produced faster gaits in less time involving minimal mechanical expertise and problem-specific knowledge.

With construction of the real robot nearing completion, future work will focus on the problem of the "reality gap" – that is, on being able to reliably transfer evolved designs into the real world. This will involve tuning the simulator to more accurately reflect reality as well as more sophisticated measures such as co-evolving the robot and the simulator and implementing learning algorithms to enable the robot to adapt to changes in itself and its environment.

References

1. Doncieux, S., Bredeche, N., Mouret, J.-B.: Exploring new horizons in evolutionary design of robots. In: Proceedings of the IEEE/RSJ International Conference on Intelligent Robots and Systems (IROS). IEEE Press (2009)
2. Harvey, I., Husbands, P., Cliff, D., Thompson, A., Jakobi, N.: Evolutionary robotics: the sussex approach. Robotics and Autonomous Systems 20, 205–224 (1997)
3. Rieffel, J., Trimmer, B., Lipson, H.: Mechanism as mind: What tensegrities and caterpillars can teach us about soft robotics. In: Artificial Life XI: Proceedings of the Eleventh International Conference on the Simulation and Synthesis of Living Systems (2008)
4. Glette, K., Hovin, M.: Evolution of Artificial Muscle-Based Robotic Locomotion in PhysX. In: IEEE/RSJ International Conference on Intelligent Robots and Systems, IROS (2010)
5. Rieffel, J., Saunders, F., Nadimpalli, S., Zhou, H., Hassoun, S., Rife, J., Trimmer, B.: Evolving soft robotic locomotion in PhysX. In: GECCO 2009: Proceedings of the 11th Annual Conference Companion on Genetic and Evolutionary Computation Conference, pp. 2499–2504. ACM, New York (2009)
6. Bongard, J.C.: Incremental Approaches to the Combined Evolution of a Robot's Body and Brain. PhD thesis, University of Zurich (2003)
7. Macinnes, I., Di Paolo, E.: Crawling out of the simulation: Evolving real robot morphologies using cheap reusable modules. In: Pollack, J., Bedau, M., Husbands, P., Ikegami, T., Watson, R. (eds.) Artificial Life IX: Proceedings of the Ninth Interational Conference on the Simulation and Synthesis of Life, pp. 94–99. MIT Press, Cambridge (2004)

8. Jakobi, N., Husbands, P., Harvey, I.: Noise and the Reality Gap: The Use of Simulation in Evolutionary Robotics. In: Morán, F., Merelo, J.J., Moreno, A., Chacon, P. (eds.) ECAL 1995. LNCS, vol. 929, pp. 704–720. Springer, Heidelberg (1995)
9. Koos, S., Mouret, J.-B., Doncieux, S.: Crossing the reality gap in evolutionary robotics by promoting transferable controllers. In: Proceedings of the 12th Annual Conference on Genetic and Evolutionary Computation, GECCO 2010, pp. 119–126. ACM (2010)
10. Bongard, J.C., Zykov, V., Lipson, H.: Resilient Machines Through Continuous Self-Modeling. Science 314(5802), 1118–1121 (2006)
11. NVIDIA, PhysX SDK, http://developer.nvidia.com/object/physx.html

Predictive Modelling of Stem Cell Differentiation and Apoptosis in *C. elegans*

Antje Beyer[1], Ralf Eberhard[2,3], Nir Piterman[4], Michael O. Hengartner[2], Alex Hajnal[2], and Jasmin Fisher[5]

[1] Department of Genetics, University of Cambridge, Cambridge, UK
[2] Institute of Molecular Life Sciences, University of Zurich, Zurich, Switzerland
[3] PhD Program in Molecular Life Sciences,
Life Science Zurich Graduate School and MD/PhD Program,
University of Zurich, Zurich, Switzerland
[4] Department of Computer Science, University of Leicester, Leicester, UK
[5] Microsoft Research, Cambridge, UK
ab704@cam.ac.uk,
{Ralf.Eberhard,Michael.Hengartner,Alex.Hajnal}@imls.uzh.ch,
Nir.Piterman@le.ac.uk, Jasmin.Fisher@microsoft.com

Abstract. The nematode *Caenorhabditis elegans* has been established as a modeling organism in biomedical research for several decades. Its hermaphrodite germ line encompasses key developmental concepts like stem cell differentiation and apoptosis; therefore it provides a good model system for these basic concepts. Here, we have extended and refined our previous computational model, which encompasses developmental landmarks and the resulting movement of germ cells along the gonadal tube. We have used the molecular dynamics (MD) framework to model the physical movement of cells due to the force arising from cell divisions. The model simulation was calibrated with experimental time and it is in accordance with experimental observations. In addition, the model provides means for testing hypotheses regarding the behaviour of mutated germ lines and the potential mechanisms causing physiological apoptosis, which are difficult to assess experimentally.

Keywords: Computational modelling, apoptosis, *C. elegans*, differentiation, molecular dynamics, germ line.

1 Introduction

Since the 1970s [1], the nematode *Caenorhabditis elegans* has been used as a modelling organism in biomedical research. In particular, its hermaphrodite germ line, shown in Fig. 1 A, provides a good model for basic developmental concepts like stem cell differentiation and apoptosis. The specific organisation of the germ line (cf. Fig. 1 B) with distinct developmental zones corresponding to specific ranges along the longitudinal axis of the germ line makes it an ideal candidate for formal analysis. Here, we have extended and refined our previous computational model [2], which encompasses these developmental landmarks and the resulting movement of germ cells

M.A. Lones et al. (Eds.): IPCAT 2012, LNCS 7223, pp. 99–104, 2012.
© Springer-Verlag Berlin Heidelberg 2012

along the gonadal tube. We have used the molecular dynamics (MD) framework to model the physical movement of cells due to the force arising from cell divisions. The model simulation was calibrated with experimental time and it is in accordance with experimental observations. In addition, the model provides means for testing hypotheses regarding the behaviour of mutated germ lines and the potential mechanisms causing physiological apoptosis, which are difficult to assess experimentally.

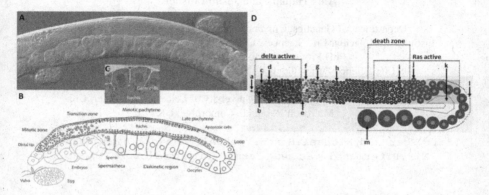

Fig. 1. The germ line of *Caenorhabditis elegans*. A: Differential interference contrast (DIC) microscopy image of the germ line shown along a plane through the centre of the gonad tube in a live animal. The head of the worm is to the right, the posterior gonad to the left of the picture. The germ cells in the meiotic pachytene region form a monolayer around a concentric inner tube (rachis), seen as a nuclei-free area in this picture. B: Schematic picture of A with developmental zones indicated. C: Cross section of the germ line by transmission electron microscopy shows germ cell monolayer with shared cytoplasm (rachis). D: Snapshot of an execution of the germ line model with zones of ligand activity and death zone indicated; a: Distal Tip Cell, b: marked cell , c: border of germ line, d: mitotic cell with highest Notch level, e: mitotic cell with Notch level between highest and 0.5 times the highest level, f: mitotic cell with Notch level between 0.5 times the highest level and 0, g: cell with GLD-1 level higher than Notch level, h: meiotic cell, i: meiotic cell that has grown to about twice its original size, j: oocyte with Ras level above threshold, k: rachis, l: border of germ line, m: fully grown oocyte.

The organisation of the germ line into distinct developmental zones is due to different signalling pathways. In the first part of the germ line, delta ligand causes the Notch level within the germ cells to peak and keeps them competent to divide. This area is superseded by a zone of GLD-1 activity whose relative levels inside the cells decide about entry into meiosis. Germ cells that have entered meiosis start to differentiate into oocytes and accumulate Ras in a specific area of the germ line with external Ras stimuli. If a cell reaches a certain level of Ras, it becomes an oocyte. The oocytes grow until they fill the diameter of the tube at the end and exit into the spermatheca to become fertilised. Corresponding to the location of the first oocytes, just before the loop in the wild-type germ line, there is a defined death area. Apoptosis is confined to this specific zone and does not happen anywhere else in the germ line.

2   Model Construction

Here, we present a dynamic computational model of the germ line of *C. elegans* based on the molecular dynamics framework. The model encompasses the previously mentioned developmental features and extends a previous model [2]. The basic component of our model is the MD movement algorithm that results in the movement of cells as they react to pressure from surrounding cells that divide, grow or are pushed themselves. That is, the major driving forces of movement in the model are cell growth and division. In this updated version of the model, we account for the special architecture of the germ line (cf. Fig. 1 A to C) with germ cells lining the outside of the tube and the shared cytoplasm in the middle of the germ line. In our two-dimensional model this three-dimensional aspect is realised using periodic boundary conditions that allow cells to interact with each other at the top and the bottom of the tube and to "jump" from top to bottom and vice versa (cf. Fig. 1 D). Furthermore, we have included the mitosis/meiosis decision due to Notch and GLD-1 in the model as follows. When a cell is in the zone of GLD-1 activity, its Notch level decays and its GLD-1 level accumulates. The cell stays in mitosis as long as the GLD-1 level is less than the Notch level. As soon as this changes, the cell enters meiosis but finishes growing until it has the size of a cell just before division (cf. Fig. 1 D). For better tractability, we changed the cell division to depend on a Gaussian distribution with average 20,000 steps and standard deviation 4,000 steps divided by 3. If we set an hour to be equivalent to 1,000 steps this assures an average time of 20 hours between divisions and a general range from 16 hours to 24 hours, which is according to the literature [3,4]. We have implemented the occurrence of apoptosis in a random, timed fashion for calibration purposes. That is, every time a cell dies the model assigns a new random time of death between 0 and 1,000 steps (~ 1h) from the current time at which another random cell within the death zone dies.

3   Results

We have calibrated our model according to literature values [3-6] and our own laboratory observations. For example, we have simulated constitutive Notch gain of function with our model, which resulted in a tumourous germ line filled with mitotic cells and no differentiating cells, which matches the experimental results [7]. This gives us confidence in the validity of the model and its appropriateness for testing further hypotheses.

The living adult germ line is difficult to assess experimentally especially since the animal moves while observing it and it is difficult to track molecular factors on a single cell level. We have used this model to perform *in silico* experiments hoping to shed new light on behaviour of mutant germ lines and on the mechanisms causing developmental apoptosis. The results of these could be used to direct experimental research. In order to evaluate the viability of these experiments we have tested a Ras loss of function mutation and the effects of increased division rate. The model predicts that a Ras loss of function mutation in the whole germ line causes a failure in

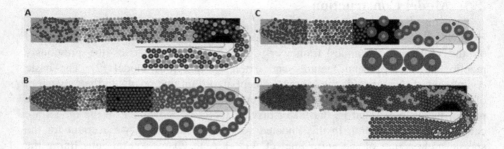

Fig. 2. Snapshots of executions of the germ line model with different mutant backgrounds. A: Ras loss of function mutation prevents differentiation into oocytes and keeps cells in meiotic pachytene. B: Ras gain of function results in death zone and oocytes to be located more distally. C: Lower division rate results in very distal oocytes and cells being more spread out. D: Higher division rate causes germ line to fill with mitotic cells and build a tumour.

A	Division	Movement	Death	Fertilisation	Variation	Look
Reference	20h	90min	2/h	2/h		
Random death	993=1h	110min	2.04/h	2.14/h	timed	good
density > 0.95	998=1h	101min	2.07/h	2.06/h	med.	med.
density, no eggs			too high			
Ras < thres.			too high			
lowest Ras	1002=1h	117min	1.97/h	2.34/h	timed	med.
highest Ras	995=1h	106min	1.98/h	2.2/h	timed	med.
size > 1.329123	992=1h	107min	1.99/h	2.25/h	good	good
smallest size	1001=1h	118min	2/h	2.23/h	timed	med.
biggest size	997=1h	117min	2/h	2.2/h	timed	med.
time in zone > 16000	1002=1h	94min	2.18/h	1.93/h	bad	med.
dev. time < thres.			too high			
dev. time > 67500	991=1h	100min	2.01/h	2.11/h	med.	med.
dev. time > 64500, no eggs	993=1h	110min	2/h	2.29/h	med.	good

Fig. 3. *In silico* experiments on possible apoptosis mechanisms. A: Table showing the values extracted from the tested hypotheses and the reference values for division, movement, death and fertilisation rates as well as personal evaluations of variation in death and fertilisation rates and the general look of the execution. Hypotheses shaded dark gray were not evaluable, hypotheses in light gray performed best in all respects. B: Death rate per hour variation over the whole execution of cell death if size > 1.329123. C: Fertilisation rate per hour variation over the whole execution of cell death if size > 1.329123.

differentiation. The germ line is filled with meiotic cells that do not differentiate into oocytes, as shown in Fig. 2 A. A gain of function in Ras, on the other hand, causes oocytes to be present more distally in the germ line of our model than under normal conditions (cf. Fig. 2 B). Furthermore, the death zone is located more distally. The model predicts that a division rate lower than normal results in very distal oocytes and a more spread out distribution of cells, as illustrated in Fig. 2 C. For an increase in the division rate, the model shows a tumourous germ line filled with mitotic cells (cf. Fig. 2 D). These results have not been tested experimentally yet, but they are in accordance with our expectations.

We use the model to direct our study of the mechanisms causing apoptosis. We have tested *in silico* six possible causes for apoptosis: completely random death, high density, Ras levels, size, time spent in death zone and developmental time (i.e. time since last division). The outcomes of these simulations are shown in Fig. 3 A. For lowest and highest Ras as well as smallest and biggest size, we used a timed death in the fashion of random death, where cells die at certain random steps of the execution, as previously explained. Fig. 3 A shows that, surprisingly, these timed deaths did not perform as well as the random death. Also, apoptosis caused by a high density did not perform as well as we had expected. In fact, the mechanisms performing best are random death and apoptosis caused by large cell size or a high developmental time (in both cases eggs are defined incompetent to die). Indeed, Fig. 3 B and C show that the death and fertilisation rates per hour for apoptosis caused by a large cell size don't show a lot of variation. The prediction of our model that large size is a very likely cause for apoptosis in the germ line makes a lot of sense biologically, assuming that the apoptotic germ cells work as nursing cells for the oocytes and, by dying, donate cytoplasm for their benefit. In this scenario, it seems effective to deplete a large cell rather than a small one. Since cells probably do not 'measure' their size, it can be considered as a phenotype of the "real" cause. Looking at our results, this could very well be the developmental time. Accordingly, we intend to focus our experimental efforts in the direction highlighted by the model. However, as previously stated, this could be very hard. We find that the computational model provides valuable predictions and inspiration for new avenues to explore experimentally *in vivo*.

Acknowledgments. This work was supported in part by the European Union grant FP7 PANACEA 222936 (Jasmin Fisher, Michael O. Hengartner and Alex Hajnal) and the Swiss National Science Foundation (Michael O. Hengartner). Antje Beyer is funded by Microsoft Research through its PhD Scholarship Programme.

References

1. Brenner, S.: The genetics of Caenorhabditis elegans. Genetics 77, 71–94 (1974)
2. Beyer, A., Eberhard, R., Piterman, N., Hengartner, M.O., Hajnal, A., Fisher, J.: A Dynamic Physical Model of Cell Migration, Differentiation and Apoptosis in Caenorhabditis elegans. In: Goryanin, I.I., Goryachev, A.B. (eds.) Advances in Systems Biology. Advances in Experimental Medicine and Biology, vol. 736, Part 2, pp. 211–233. Springer, New York (2012)
3. Crittenden, S.L., Leonhard, K.A., Byrd, D.T., Kimble, J.: Cellular analyses of the mitotic region in the Caenorhabditis elegans adult germ line. Molecular Biology of the Cell 17, 3051–3061 (2006)
4. Maciejowski, J., Ugel, N., Mishra, B., Isopi, M., Hubbard, E.J.A.: Quantitative analysis of germline mitosis in adult C. elegans. Developmental Biology 292, 142–151 (2006)
5. Hirsh, D., Oppenheim, D., Klass, M.: Development of the reproductive system of Caenorhabditis elegans. Developmental Biology 49, 200–219 (1976)

6. Gumienny, T.L., Lambie, E., Hartwieg, E., Horvitz, H.R., Hengartner, M.O.: Genetic control of programmed cell death in the Caenorhabditis elegans hermaphrodite germline. Development 126, 1011–1022 (1999)
7. Crittenden, S.L., Eckmann, C.R., Wang, L., Bernstein, D.S., Wickens, M., Kimble, J.: Regulation of the mitosis/meiosis decision in the Caenorhabditis elegans germline. Philosophical Transactions of the Royal Society of London. Series B, Biological Sciences 358, 1359–1362 (2003)

Criticality of Spatiotemporal Dynamics in Contact Mediated Pattern Formation

Nicholas S. Flann[1], Hamid Mohamadlou[1], and Gregory J. Podgorski[2,*]

[1] Department of Computer Science,
[2] Department of Biology,
Utah State University, UT 84322, U.S.A.
{nick.flann,gregory.podgorski}@usu.com

Abstract. The tissues of multicellular organisms are made of differentiated cells arranged in organized patterns. This organization emerges during development from the coupling of dynamic intra- and intercellular regulatory networks. This work applies the methods of information theory to understand how regulatory network structure within and between cells relates to the complexity of spatial patterns that emerge as a consequence of network operation. A computational study was performed in which undifferentiated cells were arranged in a two dimensional lattice, with gene expression in each cell regulated by an identical intracellular randomly generated Boolean network. Cell-cell contact signalling between embryonic cells is modeled as coupling among intracellular networks so that gene expression in one cell can influence the expression of genes in adjacent cells. In this system, the initially identical cells differentiate and form patterns of different cell types. The complexity of network structure, temporal dynamics and spatial organization is quantified through the Kolmogorov-based measures of normalized compression distance and set complexity. Results over sets of random networks from ordered, critical and chaotic domains demonstrate that: (1) Ordered and critical intracellular networks tend to create the most complex intercellular communication networks and the most information-dense patterns; (2) signalling configurations where cell-to-cell communication is non-directional mostly produce simple patterns irrespective of the internal network domain; and (3) directional signalling configurations, similar to those that function in planar cell polarity, produce the most complex patterns when the intracellular networks are non-chaotic.

1 Introduction

Multicellular organisms exhibit an incredible variety of cellular patterns, for instance, those in *Drosophila* illustrated in Figure 1. These patterns arise during development and are a consequence of genetic regulatory networks (GRNs) that operate within cells in response to communication between cells [19, 20]. An

---

* Supported by the Luxembourg Centre for Systems Biomedicine and the University of Luxembourg.

M.A. Lones et al. (Eds.): IPCAT 2012, LNCS 7223, pp. 105–116, 2012.

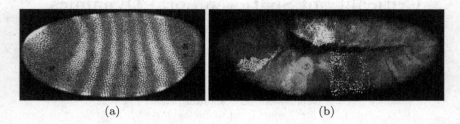

(a)   (b)

Fig. 1. Example of pattern formation in *Drosophila*. Panel (a) shows expression of a pair-rule gene expression in the cellular blastoderm (Courtesy of Langeland, S. Paddock, and S. Carroll, HHMI) and panel (b) shows the expression of Hox genes at the extended germ band stage. (Courtesy ofDave Kosman, UCSD).

interesting question to explore is the relationship between the structure of GRNs and intercellular communication networks that connect cells and the complexity of cellular patterns that can emerge from the operation of these networks.

Although details are not known for how multicellularity evolved from a unicellular ancestor in any lineage, complex multicellularity almost certainly arose from the co-option of GRNs and intercellular communication systems that existed in single-celled organisms [18]. While the actual path of evolution to complex muticellularity may never be known, the paths open to evolution can potentially be explored and understood through computational studies. This is a long term goal of the investigations reported here.

Evidence suggests that living processes lie "on the edge of chaos" in that evolution has selected for biological systems that maximally retain information yet allow the system to evolve [15, 16, 22]. Systems operate in three complexity domains: ordered, critical and chaotic. Ordered systems are robust in that they damp perturbations to retain information, but they have limited the potential for change. Chaotic systems magnify perturbations and lose information, rendering them unsuitable for homeostatic living systems; in fact they are implicated in diseases like cancer. Critical systems, which operate in the narrow region between order and chaos, are the most information dense in both structure and dynamics. This work focuses on how the information content of cellular patterns is influenced by the complexity domain of intracellular genetic regulatory networks and the nature of cell-cell signalling.

2   Methods

An empirical study was performed with a simulated embryonic epithelium consisting of a grid of undifferentiated cells, each containing identical Boolean networks to model the genetic regulatory network. Cell-cell communication was modeled by linking the output of a Boolean function to the input of the genetic regulatory network controlling one or more adjacent cells. The complexity

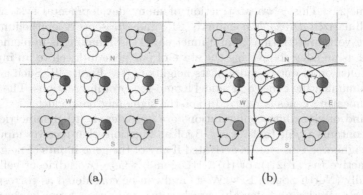

(a)   (b)

Fig. 2. Model used in this work. Each box is a cell within the epithelium containing an intracellular Boolean network that is identical to all other cells within the epithelium. A genetic regulatory network is represented as a graph where nodes are Boolean functions (representing a gene regulatory function) and edges denote an interaction between the output of one function and the input of another (a different regulatory gene). In (a) there is no signalling between cells; (b) illustrates orthogonal contact-mediated communication where one gene regulates the expression of another gene in an adjacent cell. Red nodes represent communicating genes, white are intracellular genes.

domain of the network, its temporal dynamics and resultant pattern were quantified using the information theoretic measure called set complexity [10]. Empirical studies were performed over ensembles of randomly generated Boolean networks and the results compiled. Each step is defined in detail below.

2.1   Regulatory Network Models

Boolean networks [13] represent either expressed or non-expressed genes that are regulated by other genes using logic functions. In this work, sets of random intracellular Boolean networks were generated by randomly interconnecting a varying number of nodes within one cell then instantiating each regulatory node with a randomly generated logic function.

To produce networks of different complexity domains, the number of inputs to each Boolean function (node in the network) is set according to $s = 2kp(1-p)$ where s is the sensitivity of the network to perturbations in gene values, p is the probability of the output of each Boolean function being a 1, and k is the count of inputs to each Boolean function [8]. When $s = 1$ a single bit change is on average propagated to one other node and the network is in the critical domain. In an ordered network, $s < 1$ and perturbations tend to die out, while in a chaotic network, $s > 1$ and perturbations tend to grow. In this work, p was fixed at 0.5 and k was changed to create networks of different complexity domains: $k = 1$ for ordered, $k = 2$ for critical, and $k = 3$ for chaotic.

To study multicellular pattern formation, a two dimensional lattice of cells containing identical GRNs was employed as a simple model of an embryonic epithelium [24]. This is an abstraction of many developmental systems, such as the cellularized *Drosophila* embryo [21], and the sensory epithelium of the developing vertebrate retina [9] and inner ear [11]. Signalling is implemented in the model as an edge connecting the state of one gene in a cell to an input of a Boolean function of one or more of its neighbors (see Figure 2). Such genes are called communicating genes and model ligand-receptor interactions. The number of communicating genes is referred to as the signalling bandwidth.

Two kind of signalling configurations are considered: (a) Symmetric, where each cell contains a gene (output of a Boolean function) that receives inputs from all four neighbors. This gene is activated if any of the gene inputs of neighboring cells are active (as in [27]); or (b) Orthogonal, where two adjacent cells signal directionally (North-South, East-West) and can be considered as corresponding to anterior-posterior and dorsal-ventral embryonic axes. A mechanism to autonomously generate this intercellular directionality via a morphogen gradient has been elegantly demonstrated in [17]. This is implemented in the Boolean network by connecting an output function of the originating cell to the input function of the destination cell.

Since this work focusses on the self organization of patterns, the state of each intracellular network is initialized randomly by setting the activation of each gene to on or off with equal probability. To simulate the emergence of patterns over the modeled epithelium, the state of the system (represented as the value of each gene in each network) is clocked synchronously until either a steady state or the maximum number of updates is reached. Synchronous updating was employed, rather than asynchronous updating [26], because it has proved useful in modeling and analysis applications, for instance in the identification of criticality in eukaryotic cell dynamics [25] and in the accurate modeling of *Drosophila* segment formation [1; 4]. Additionally, synchronous updating simplifies the identification of common attractors among cells, needed in pattern encoding and complexity domain determination.

A steady state of an intracellular network represents an attractor of the network [13] and may be cyclic. To detect that a cell is in a steady state, at each time period the current state of the intracellular network is compared to all its previous states. If a single match is found, an attractor has been reached since the updates are deterministic. If no cycle is detected within the maximum number of steps, the cell is considered to be in a chaotic state. Attractors represent terminally differentiated cell types [12], and the trajectory to an attractor represents the process of cell differentiation. In this work, if the state of two cells converge to the same attractor, even if they are out of phase, then they are considered as the same cell type.

2.2   Information Complexity

Set complexity [10] can be used to measure the information content of regulatory networks, their temporal dynamics and the spatial patterns produced. By measuring information content, set complexity can distinguish between critical systems that encode maximal information, and ordered and chaotic systems that encode low information. Set complexity (referred to as Ψ) applies Kolmogorov's intrinsic complexity to quantify contextual information in a set of objects by discounting pairs of objects that are randomly related or redundant. Set complexity is independent of any specific application, so long as each object in the set can be encoded as a string.

The Kolmogorov complexity of two strings is the length of the shortest algorithm that can transform one string to the other. Exact computation is undecidable, but minimum algorithm length can be approximated by the normalized compression distance (NCD) [6, 7]. NCD is defined below, where x and y are strings, xy is the concatenation of x and y and $C(x)$ is the compression size of x:

$$NCD(x,y) = \frac{C(xy) - \min(C(x), C(y))}{\max(C(x), C(y))}, 0.0 \leq NCD(x,y) \leq 1.0 \qquad (1)$$

NCD is a measure of the information content of the two strings [2]. Consider the following cases, where dissimilar strings could be a random string paired with a string of a single repeated character:

Similar strings: $x \simeq y$, $C(xy) \simeq C(x) \simeq C(y)$ then $NCD(x,y) \simeq 0$

Random strings: $x \neq y$, $C(xy) \simeq C(x) + C(y)$ then $NCD(x,y) \simeq 1$

Dissimilar strings: $x \neq y$, $C(xy) \simeq C(x), C(y) = 0$ then $NCD(x,y) \simeq 1$

To ensure accurate measurement of compression length the block size of the compressor must be greater than the string length. Here we used the bzip2 compression algorithm with a block size of 900 Kbytes [5].

Then set complexity of a set of n strings $S = \{x_1, \ldots, x_n\}$ is defined:

$$\Psi(S) = \sum_{x_j \in S} C(x_j) \frac{1}{n(n-1)} \sum_{x_j \neq x_k} d_{jk}(1 - d_{jk}) \qquad (2)$$

where $d_{ij} = NCD(x_i, x_j)$. The distance d_{ij} measures the mutual information between the two strings and is maximized when $NCD(x_i, x_j) = 0.5$ and minimized when $NCD(x_i, x_j) = 0.0$ or $NCD(x_i, x_j) = 1.0$. When strings in the set are similar, $\Psi(S) \simeq 0$ indicating the set belongs to the ordered domain and contains little information. Chaotic systems generate strings that appear random and then $\Psi(S)$ is minimized, but not zero because of the $C(x_j)$ multiplicative term. In [10] it is shown that $\Psi(S)$ is maximized when the set of strings describe a critical system.

2.3   Network, Dynamics and Pattern Encoding

To compute the set complexity, each random network constructed (denoted Ψ_n), its temporal dynamics (denoted Ψ_t) and the spatial pattern produced (denoted Ψ_p) must be encoded as a string by a one-to-one mapping so that no information is lost. Studies in [23] suggest that *NCD* and Ψ are in general insensitive to the specific encoding methods employed so long as the compression methods are effective. Let n be the number of Boolean functions in each intracellular network, k be the number of input connections of each function and m^2 be the total number of cells in the pattern (for a square pattern of $m \times m$). The following mappings were employed:

Network: The method used is described in [3] (supplementary materials). Here the complete intercellular network is represented as a directional connectivity matrix with side m^2nk where each input of each Boolean function is a unique node. The matrix is then represented in row-order and encoded as a string. Each Boolean function is encoded by 2^k 1's or 0's, one for each row in the function table, and the k identifiers of its input nodes used in the connectivity matrix. The two strings are then concatenated.

Temporal dynamics: To simulate pattern formation, each network is executed for 300 time steps with a "burn in" period of 100 steps [23], which is ignored in the analysis of the dynamics. The 2D space-time matrix of the network state trajectory with size $200c^2n$ is then encoded as a row-order string of 1's and 0's.

Spatial pattern: At the completion of the forward simulation of the network, the dynamics of each intracellular network is analyzed to identify cyclic attractors by searching for repeating states. Then each cell is assigned a cell type ID by performing $100m^2$ comparisons where matching attractors are assigned the same type (irrespective of phase). The string is then a row-order concatenation of each cell's type ID.

3   Experimental Study

In this study the number of Boolean functions in each intracellular network is fixed at eight and the pattern is fixed at a 20 by 20 square. These values represent a balance between computational feasibility and realism. The entire intercellular network contains 3200 Boolean functions. To simulate the activity of the network, each gene in each cell is randomly assigned a value and stepped forward 300 iterations as described in Section 2.3.

With three complexity domains of intracellular networks (ordered, critical and chaotic), two communication configurations (symmetric and orthogonal), and nine possible bandwidths (zero through eight) there are 54 experimental conditions. For each condition, 40 random networks were constructed and each executed 10 times from a distinct random initial state. For each run, the specific network, temporal dynamics and resulting spatial pattern were encoded into strings as described in Section 2.3 and stored in separate folders. Given these parameters, the string size of the network is $3200k2^k$ characters; the string size of the dynamics is $64000k$ characters; and the string size of the network is $200m$ characters, where m is the number of unique attractors. Additionally, each spatial pattern was recorded as an image, examples of which are provided in Figure 3.

Ψ_p	BW.	comm.	domain	Example Patterns				
9.38	6	sym.	ordered					
18.85	5	orth.	ordered					
19.92	5	sym.	critical					
26.94	6	orth.	critical					
6.72	7	sym.	chaotic					
12.26	8	orth	chaotic					

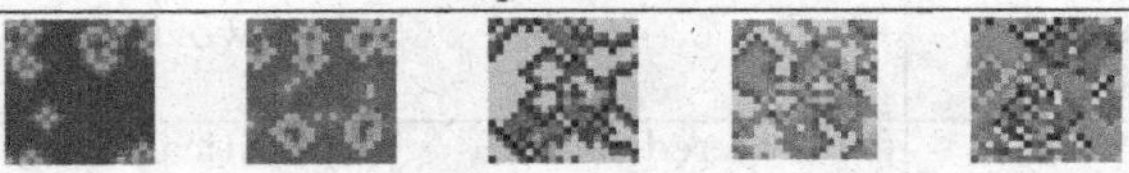

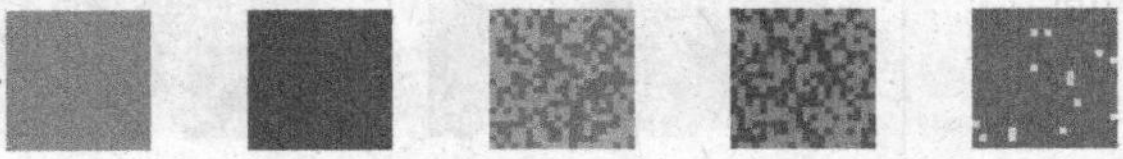

Fig. 3. Examples of human selected patterns from result sets showing their Ψ_p value, the bandwidth (BW is number of communicating genes), the intercellular signalling configuration (orth; is orthogonal, sym; is symmetrical) and the complexity domain (ordered, critical, chaotic) of the intracellular network. Each cell in the pattern is colored according to its attractor (same attractor, same color). Patterns are ordered left to right by increasing compression size.

Results presented in Section 4 were computed for each experimental condition above. Four hundred network executions were preformed for each experimental condition. For every execution of a network its dynamics and pattern were encoded as strings and stored. For each of these string sets, 2000 *NCD* values were computed by randomly sampling string pairs. Not all pairs of were considered because the total number of *NCD* values grows as the square of the string set cardinality, see Equation (1). These results are presented in Figure 4 for the patterns only; distributions for dynamics and networks are not shown. Next, Ψ was computed for the network, dynamics and pattern string sets for each of the 54 experimental conditions. Ψ was estimated from sampling by averaging 100 distinct set complexity computations, each determined from a random sampling of 10 *NCD* values. Sampling was used since the run time of set complexity grows as the square of *NCD* set cardinality, see Equation (2).

4   Results and Discussion

Fig. 3 illustrates six sample pattern sets from the 54 possible experimental conditions (network complexity domain, signalling configuration, and bandwidth of

Comm. Type	Network Domain		

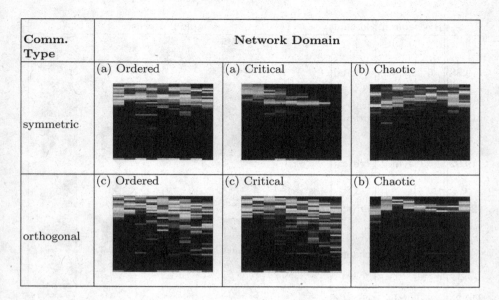

Fig. 4. The distribution of *NCD* values between pattern pairs as a function of the signalling bandwidth (number of communicating genes) from 0 to 8 along the horizontal axis. The vertical axis for each plot is the *NCD* range from 0.0 (bottom) to 1.0 (top). High probability is red, low probability is blue. The (a), (b) and (c) refer to the equivalent classes discussed in Section 4.1.

communication) described above. Within each experimental condition a diversity of patterns is produced dependent upon the specific intracellular network topology, Boolean functions generated, the specific intercellular network topology and the random initial conditions. However, commonalities are visually apparent within each experimental condition. These commonalities were quantified by computing the Ψ_p values of the pattern sets. For example, the first row of Figure 3 shows simple patterns of regular patches set on a uniform background, with a low Ψ_p of 9.38. In contrast, the fourth row shows complex diagonal repeating elements with varying periodicity and high Ψ_p of 26.94. Symmetric signalling tends to produce patterns that contain contiguous regions and maze-like interfaces that have low set complexity. Orthogonal signalling tends to produce repeating regular pattern elements that have high set complexity.

4.1   *NCD* Analysis of Pattern Sets

Figure 4 illustrates the distributions of *NCD* values for pairs taken from sets of patterns generated under symmetrical or orthogonal signalling, and ordered, critical and chaotic networks. Recall that the *NCD* of two identical patterns is 0.0 and two random patterns is 1.0. Also that the mutual information between two strings x_i, x_j is maximized when $NCD(x_i, x_j) \simeq 0.5$ and minimized when

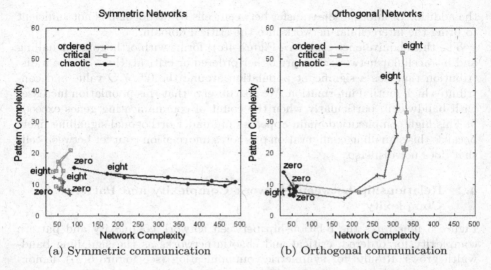

Fig. 5. The relationship between network complexity and the subsequent pattern complexity for the two signalling configurations. Each line shows the trajectory as the signalling bandwidth increases from zero to eight.

$NCD(x_i, x_j) = 0.0$ or $NCD(x_i, x_j) = 1.0$. The distributions of NCD provide insights into the diversity of patterns produced and why certain network configurations produce higher set complexity and others provide low set complexity. If many pairs of patterns have an NCD distance near 0.5, then Ψ_p will be high because of the sum of mutual information $NCD(x_i, x_j)(1 - NCD(x_i, x_j))$ will be high. If most pairs of patterns drawn from the set are identical or random, then NCD will be bimodal distributed at 0.0 and 1.0, and the set will have low Ψ_p.

Figure 4 shows that the six network configurations can be placed into three equivalent classes based on their distributions. The first equivalence class (see Figure 4(a)) is symmetric signalling with ordered or critical networks, which has a clear bimodal distribution of NCD with a maxima at 0.0 and near 1.0. The 0.0 maxima is caused by most cells following a trajectory to the same attractor because either the intracellular network only has a few attractors, or the connectivity over-constrains the attractor landscape. The 1.0 maxima is caused by cells quickly converging to distinct attractors forming a pattern with little or no spatial organization. Information transfer is decays between the cells because of the damping effect of the symmetric connections.

The second equivalence class (see Figure 4(b)) forms when the intracellular networks are chaotic, irrespective of the signalling configuration. Here, all NCD values are near 1.0 because the patterns are either disordered or complex with many errors (as illustrated in Figure 3). It is well known [14], that chaotic networks produce dynamics with some attractors of long or unlimited time steps, so that some of the cells in the pattern appear as a unique type. Significantly, the addition of information transfer between cells by orthogonal is not sufficient to bring the intercellular network into the critical domain.

The third equivalence class (see Figure 4(c)) forms with orthogonal signalling and intracellular network that are either ordered or critical. Here, we see a distribution that has a significant population around the $0.5NCD$ value and contributes high mutual information. We also observe that this population increases with bandwidth, particularly when the count of communicating genes exceeds 5. This high complexity domain appears only under orthogonal signalling likely because this signalling configuration promotes information transfer between cells that does not disperse.

4.2   Relationship between Network Complexity and Pattern Complexity

Figure 5 illustrates the relationship between network complexity and pattern complexity for ordered, critical and chaotic networks as the signalling bandwidth grows. Results for symmetric communication (see Figure 5(a)) demonstrate that this configuration is sufficient to generate low complexity patterns in the simulated epithelium. The complexity domain of the network has little effect on pattern complexity with all three domains producing a narrow range of low complexity patterns. Interestingly, increases in signalling bandwidth have little effect on network complexity for ordered or critical networks. However, when intracellular networks are chaotic, increasing bandwidth produces high network complexity with no increase in pattern complexity. The cause of this behavior is unknown and open to further study.

With the introduction of orthogonal directionality (see Figure 5(b)) we see significant changes in pattern complexity. For critical and ordered networks, as signalling bandwidth grows from 1 to near half the intracellular genes, the network complexity maximizes but has little effect on the pattern complexity. Beyond this point, increases in signalling bandwidth maintain network complexity but significantly increase pattern complexity to a maximum when every intracellular gene is communicating. Chaotic networks using orthogonal signalling cannot develop complex patterns or complex networks irrespective of communication bandwidth.

5   Summary

This work has explored the potential of ordered, critical or chaotic genetic regulatory networks to create complex patterns in an simulated field of embryonic cells. The impact of the transition from autonomous cells to cells that communicate by contact-mediated signalling was examined as the number of signalling connections increase. An information theoretic measure was used to evaluate the information content of the originating networks and the emergent cellular patterns. The most complex patterns emerge from ordered and critical networks that communicate directionally. When cells communicate with all neighbors isotropically, only simple, low information patterns emerge. Low information patterns also emerge from chaotic networks regardless of the signalling bandwidth or configuration.

References

1. Albert, R., Othmer, H.G.: The topology of the regulatory interactions predicts the expression pattern of the segment polarity genes in Drosophila melanogaster (2003), http://citeseerx.ist.psu.edu/viewdoc/summary?doi=10.1.1.13.3370
2. Án, M.C., Alfonseca, M., Ortega, A.: Common pitfalls using normalized compression distance: what to watch out for in a compressor. Communications in Information and Systems 5, 367–384 (2005), http://citeseerx.ist.psu.edu/viewdoc/summary?doi=10.1.1.104.9265
3. Balleza, E., Alvarez-Buylla, E.R., Chaos, A., Kauffman, S., Shmulevich, I., Aldana, M.: Critical Dynamics in Genetic Regulatory Networks: Examples from Four Kingdoms. PLoS ONE 3(6), e2456+ (2008), http://dx.doi.org/10.1371/journal.pone.0002456, doi:10.1371/journal.pone.0002456
4. Bodnar, J.W.: Programming the Drosophila embryo. Journal of Theoretical Biology 188(4), 391–445 (1997), http://dx.doi.org/10.1006/jtbi.1996.0328
5. Burrows, M., Wheeler, D.J., Burrows, M., Wheeler, D.J.: A block-sorting lossless data compression algorithm (1994), http://citeseerx.ist.psu.edu/viewdoc/summary?doi=10.1.1.121.6177
6. Chen, X., Francia, B., Li, M., McKinnon, B., Seker, A.: Shared Information and Program Plagiarism Detection. IEEE Transactions on Information Theory 50(7), 1545–1551 (2004), http://dx.doi.org/10.1109/TIT.2004.830793
7. Cilibrasi, R., Vitanyi, P.M.B.: Clustering by compression. IEEE Transactions on Information Theory 51(4), 1523–1545 (2005), http://dx.doi.org/10.1109/TIT.2005.844059
8. Derrida, B., Pomeau, Y.: Random Networks of Automata: A Simple Annealed Approximation. EPL (Europhysics Letters) 1(2), 45–49 (1986), http://dx.doi.org/10.1209/0295-5075/1/2/001
9. Eglen, S.J., Willshaw, D.J.: Influence of cell fate mechanisms upon retinal mosaic formation: a modelling study. Development 129(23), 5399–5408 (2002), http://view.ncbi.nlm.nih.gov/pubmed/12403711
10. Galas, D.J., Nykter, M., Carter, G.W., Price, N.D., Shmulevich, I.: Biological Information as Set-Based Complexity. IEEE Transactions on Information Theory 56(2), 667–677 (2010), http://dx.doi.org/10.1109/TIT.2009.2037046
11. Goodyear, R., Richardson, G.: Pattern formation in the basilar papilla: evidence for cell rearrangement. The Journal of Neuroscience: the Official Journal of the Society for Neuroscience 17(16), 6289–6301 (1997), http://view.ncbi.nlm.nih.gov/pubmed/9236239
12. Huang, S., Eichler, G., Yam, Y.B., Ingber, D.E.: Cell Fates as High-Dimensional Attractor States of a Complex Gene Regulatory Network. Physical Review Letters 94(12), 128701+ (2005), http://dx.doi.org/10.1103/PhysRevLett.94.128701
13. Kauffman, S.A.: Metabolic stability and epigenesis in randomly constructed genetic nets. Journal of Theoretical Biology 22(3), 437–467 (1969), http://view.ncbi.nlm.nih.gov/pubmed/5803332
14. Kauffman, S.A.: The Origins of Order: Self-Organization and Selection in Evolution, 1st edn. Oxford University Press, USA (1993), http://www.worldcat.org/isbn/0195079515

15. Kauffman, S.A., Johnsen, S.: Coevolution to the edge of chaos: Coupled fitness landscapes, poised states, and coevolutionary avalanches. Journal of Theoretical Biology 149(4), 467–505 (1991),
http://dx.doi.org/10.1016/S0022-5193(05)80094-3

16. Kitzbichler, M.G., Smith, M.L., Christensen, S.R., Bullmore, E.: Broadband Criticality of Human Brain Network Synchronization. PLoS Comput. Biol. 5(3), e1000314+ (2009),
http://dx.doi.org/10.1371/journal.pcbi.1000314

17. Knabe, J.F., Nehaniv, C.L., Schilstra, M.J.: Evolution and morphogenesis of differentiated multicellular organisms: Autonomously generated diffusion gradients for positional information. In: Bullock, S., Noble, J., Watson, R., Bedau, M.A. (eds.) Artificial Life XI: Proceedings of the Eleventh International Conference on the Simulation and Synthesis of Living Systems, pp. 321–328. MIT Press (2008),
http://panmental.de/ALifeXIflag

18. Knoll, A.H.: The Multiple Origins of Complex Multicellularity. Annual Review of Earth and Planetary Sciences 39(1), 217–239 (2011),
http://dx.doi.org/10.1146/annurev.earth.031208.100209

19. Lander, A.D.: Morpheus Unbound: Reimagining the Morphogen Gradient. Cell 128(2), 245–256 (2007),
http://dx.doi.org/10.1016/j.cell.2007.01.004

20. Lander, A.D.: Pattern, Growth, and Control. Cell 144(6), 955–969 (2011),
http://dx.doi.org/10.1016/j.cell.2011.03.009

21. Mazumdar, A., Mazumdar, M.: How one becomes many: blastoderm cellularization in Drosophila melanogaster. BioEssays: News and Reviews in Molecular, Cellular and Developmental Biology 24(11), 1012–1022 (2002),
http://dx.doi.org/10.1002/bies.10184

22. Mitchell, M., Hraber, P., Crutchfield, J.P.: Revisiting the Edge of Chaos: Evolving Cellular Automata to Perform Computations (March 1993),
http://arxiv.org/abs/adap-org/9303003

23. Nykter, M., Price, N.D., Larjo, A., Aho, T., Kauffman, S.A., Harja, O.Y., Shmulevich, I.: Critical Networks Exhibit Maximal Information Diversity in Structure-Dynamics Relationships. Physical Review Letters 100(5), 058702+ (2008),
http://dx.doi.org/10.1103/PhysRevLett.100.058702

24. Serra, R., Villani, M., Damiani, C., Graudenzi, A., Colacci, A.: The Diffusion of Perturbations in a Model of Coupled Random Boolean Networks. In: Umeo, H., Morishita, S., Nishinari, K., Komatsuzaki, T., Bandini, S. (eds.) ACRI 2008. LNCS, vol. 5191, pp. 315–322. Springer, Heidelberg (2008),
http://dx.doi.org/10.1007/978-3-540-79992-4_40

25. Shmulevich, I., Kauffman, S.A., Aldana, M.: Eukaryotic cells are dynamically ordered or critical but not chaotic. Proceedings of the National Academy of Sciences of the United States of America 102(38), 13439–13444 (2005),
http://dx.doi.org/10.1073/pnas.0506771102

26. Thomas, R.: Regulatory networks seen as asynchronous automata: A logical description. Journal of Theoretical Biology 153(1), 1–23 (1991),
http://dx.doi.org/10.1016/S0022-5193(05)80350-9

27. Villani, M., Serra, R., Ingrami, P., Kauffman, S.A.: Coupled Random Boolean Network Forming an Artificial Tissue. In: El Yacoubi, S., Chopard, B., Bandini, S. (eds.) ACRI 2006. LNCS, vol. 4173, pp. 548–556. Springer, Heidelberg (2006),
http://dx.doi.org/10.1007/11861201_63

The Vasopressin System – Asynchronous Burst Firing as a Signal Encoding Mechanism

Duncan J. MacGregor, Tom F. Clayton, and Gareth Leng

Centre for Integrative Physiology, University of Edinburgh, UK
Gareth.Leng@ed.ac.uk

Abstract. The task of the vasopressin system is homeostasis, a type of process which is fundamental to the brain's regulation of the body, exists in many different systems, and is vital to health and survival. Many illnesses are related to the dysfunction of homeostatic systems, including high blood pressure, obesity and diabetes. Beyond the vasopressin system's own importance, in regulating osmotic pressure, it presents an accessible model where we can learn how the features of homeostatic systems generally relate to their function, and potentially develop treatments. The vasopressin system is an important model system in neuroscience because it presents an accessible system in which to investigate the function and importance of, for example, dendritic release and burst firing, both of which are found in many systems of the brain. We have only recently begun to understand the contribution of dendritic release to neuronal function and information processing. Burst firing has most commonly been associated with rhythm generation; in this system it clearly plays a different role, still to be understood fully.

Keywords: vasopressin, phasic firing, dendritic release, modelling.

1 Introduction

We now recognise that rather than just simple integrators and relays of activity, most neurons have complex pattern generating properties. An important question in contemporary neuroscience is how these properties contribute to information processing [6, 27]. In particular, many neurons generate "bursting" patterns of electrical activity, arising either through intrinsic mechanisms, or via network interactions. Some of these contribute to generating physiological rhythms (such as the respiratory rhythm, [8]); neurons synchronising across a network to generate an emergent rhythm. In others, single synchronised bursts are essential to the physiological output, such as oxytocin cells driving the periodic milk let-down during suckling [29]. However, some neurons, like the vasopressin cells of the hypothalamus, display bursting activity (Fig. 1), but fire *asynchronously* [19] - so what is *this* bursting behaviour for? We know that their bursts are efficient for stimulus-secretion coupling, optimising secretion per spike, but it is not clear why this is important - many neurons fire spontaneously at much higher rates than vasopressin cells. Moreover, bursting is efficient in these cells only because

M.A. Lones et al. (Eds.): IPCAT 2012, LNCS 7223, pp. 117–130, 2012.

of particular properties of their axon terminals, indicating that these properties have co-evolved with the bursting behaviour; suggesting that bursting is important for other reasons. Because bursting is such a widespread feature in the CNS, arising in many different ways, we believe it is important to understand exactly what advantages it offers for information processing.

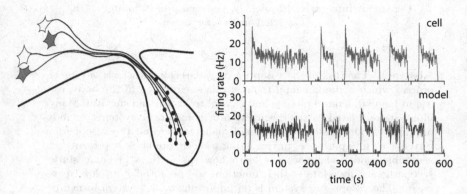

Fig. 1. Vasopressin cells project to the posterior pituitary. In response to osmotic input, cells fire in a distinct phasic bursting pattern. We can closely match this behaviour with a relatively simple single cell model.

Vasopressin and its control of osmotic pressure is a relatively simple and very well studied system. It presents an unusually strong opportunity to be able to relate information processing properties of cells to their physiological function as part of a system. We are currently attempting to apply a modelling and complex systems approach, in order to test specific hypothesis about the adaptive value of its particular features (heterogeneity, bistability, autocrine and paracrine communication mechanisms). We will test these features ultimately by expressing the physiological function of the system in terms of a defined control task, integrating neuronal modelling into a physiological systems model. The project is defined in three parts:

1. Build a single neuron model, including spike firing, vasopressin secretion and intercellular communication mechanisms.
2. Duplicate the model to build a network. Evaluate input/output characteristics and study the effects of varied assumptions about communication.
3. Build a closed-loop system model and use this to test varied network models, comparing their performance in matching experimental data, and systematically evaluate network performance, robustness and efficiency.

The first details of these components are presented elsewhere [7]. What we seek to develop here is the rationale and strategy behind the work. Though these parts require initially to be developed in sequence, they will continue to be refined in parallel, better informed by their role and behaviour as part of a system.

2   Background: Homeostatic Role of Vasopressin

Vasopressin is made by neurons of the supraoptic and paraventricular nuclei of the hypothalamus, and is secreted into the blood from axonal terminals at the posterior pituitary. This is a very important model system in neuroscience for many reasons, including the large size and accessibility of the neurons, and the fact that, because these neurons secrete their products at measurable amounts into the systemic circulation, their electrical activity can be directly related to secretion and physiological function. These cells use cell volume modulated stretch-sensitive channels to respond to osmotic pressure, and also receive synaptic input from other osmosensitive neurons [4].

Vasopressin cells display relatively long bursts and silences. They are bistable oscillators; small perturbations can "flip" a neuron from either state (burst or silence) to the other, because their intrinsic activity-dependent conductances can either stop or start a burst. This means that an asynchronously firing population of vasopressin cells can act as a low-pass filter - preserving low frequency signals while filtering out stochastic noise in their inputs [32]. While individual neurons respond erratically to acute changes, the asynchronicity means that these erratic responses are smoothed out. The vasopressin cells are also a heterogeneous population; variation in expression of membrane channels, receptors, and synaptic input, produce differing sensitivities to osmotic pressure and a wide spectrum of bursting behaviour. This heterogeneity has been preserved through evolution, suggesting either that it is an inescapable limitation, or, more interestingly, that the heterogeneity is adaptive. There are some clear adaptive consequences of heterogeneity - a population that is heterogeneous in osmotic responsiveness will have a wider dynamic range than a homogeneous population. But there are also costs; for example, a homogeneous population has a high intrinsic redundancy, so it is robust to degradation. A heterogeneous population will generally be less robust - unless the heterogeneity is not hard-wired, but arises from network self organisation. The most obvious way that heterogeneity could be self-organised would be if individual neurons cycle through phases of varying osmotic responsiveness - as we have suggested that they might [19]. The population of vasopressin neurons act together as a complex system, with multiple feedbacks acting at different levels, including autocrine signals and paracrine signalling between cells. These properties co-ordinate the vasopressin cells, presumably to optimise emergent features of system behaviour.

The vasopressin-osmotic system is part of a larger homeostatic system that regulates plasma volume and electrolyte concentration via many mechanisms (including thirst and natriuresis; see [3]). Vasopressin secretion (Fig. 2) increases linearly with osmolarity above a set point [10], and this is essential for regulation of plasma volume and osmolarity. Plasma osmolarity is normally regulated to within a few percent, so vasopressin cells, as a population, must respond reliably to a change in extracellular $[Na^+]$ of just ∼1mM - tiny compared to the fluctuations expected as the result of stochastic variations in neuronal activity. A sustained increase in osmolarity requires a sustained vasopressin response, so the vasopressin cell population must maintain their response to an unvarying input signal. Most neurons are good at responding to change, but to do this they adapt to a constant signal; vasopressin cells as a population must not adapt to sustained osmotic stimulation.

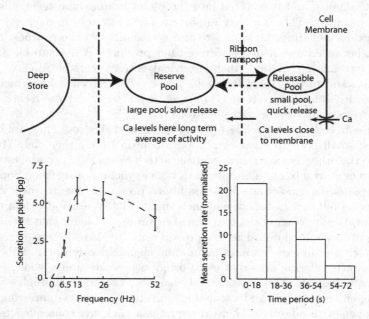

Fig. 2. Vasopressin is synthesised in the cell body and transported to the release sites through a sequence of pools, with transport and release activity driven by spike triggered Ca2+ entry, and also possibly internal stores. The lower panels show in-vitro data from [2]. Stimulus-secretion coupling is non-linearly dependent on both spike rate (lower left) and burst duration (lower right). Per spike secretion is optimal at ∼13Hz, initially showing facilitation, before being limited by the releasable pool. The secretion rate also declines during prolonged bursts as the reserve pool is depleted.

At normal osmotic pressures, the cells fire slowly; each secretes just 1-2 vesicles/s, but this is enough to maintain normal circulating concentration of ∼1pg/ml (see [20] for details). As osmotic input increases, the cells enter a bistable phasic firing mode [32], consisting of alternating bursts and silences. Each burst typically lasts for 20-60s at 4-10 spikes/s. Secretion is facilitated by high frequency spiking, but fatigues within about 20s; this fatigue is reversed after 20-30s of quiescence; thus a phasic firing pattern optimises secretion per spike [2]. A burst of ∼400 spikes in one vasopressin cell, releases about one vesicle from each of its ∼2,000 nerve endings. However, with chronic stimulation, the stores of vasopressin are progressively depleted; if rats are given 2% NaCl to drink, then stores decline to ∼15% of control values over 12 days, despite a massive increase in synthesis [17]. This decline reflects the delay between increasing the rate of synthesis and replenishment of the stores. At any particular time, hormone secretion in response to a given stimulus is proportional to the size of the store [13]; thus, during progressive dehydration, spike activity becomes less and less effective at secreting vasopressin.

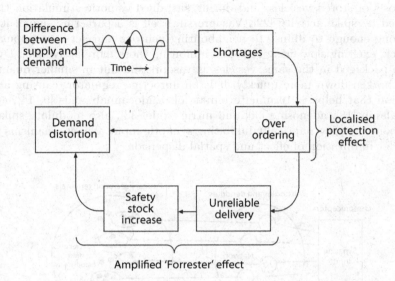

Fig. 3. The 'Forrester flywheel' summarizes common problems in supply chains [35]. In business, stock levels incur space and wastage costs so must be kept low; but if stocks run out, delays in restocking mean lost sales. In response to fluctuations in demand (synaptic input) the business can alter manufacture (synthesis) and moderate supply (secretion) by moderating price levels (stimulus-secretion coupling). Management needs to link manufacturing (synthesis) to sales (secretion); to link stock (stores) levels to a given variability of demand (expected variability of physiological challenge) for given delays in the system; and to ration supply by raising prices. The business must minimise the risks of losses associated with either running out of stock (hypernatraemia) or overstocking.

The larger homeostatic system, of which the vasopressin system is part, must regulate plasma osmolarity and volume within strict tolerance. Both hyper- and hyponatraemia are life-threatening outside critical limits. We propose that the utility of this system should be judged not by how accurately it maintains normal osmolarity, but by how well it can prolong survival – i.e., when subject to chronic challenge, how long will it maintain osmolarity within tolerable bounds? This is a novel way of understanding the vasopressin system; it expresses its physiological function in terms of a defined control task, and in so doing it enables systematic, objective study of its performance, and systematic assessment of the utility of each of the various features of the vasopressin system.

This task is not trivial, and is in fact analogous to a problem in supply chain management that has considerable economic significance (Fig. 3).

3   Network Structure and Regulation

Vasopressin cells are not synaptically interconnected, but communicate via dendritic release of several substances (Fig. 4). This requires Ca2+ dependent exocytosis of stored vesicles, and during sustained osmotic stimulation this is triggered by spike activity [22]. Vasopressin itself is a paracrine signal; it survives long enough to diffuse to neighbouring neurons and act as a population feedback, exciting slow firing cells and inhibiting those more active [12]. Dynorphin is packaged in the same vesicles as vasopressin, but in smaller quantities and is broken down more quickly; it is an autocrine regulator, causing a slow inhibition that helps to terminate bursts [5]. Endocannabinoids [9, 14], apelin [21], galanin [18], adenosine [30] and nitric oxide [33], also modulate spike activity, some by presynaptically inhibiting synaptic input. The mechanisms vary, and differ in duration of effect and spatial dispersion.

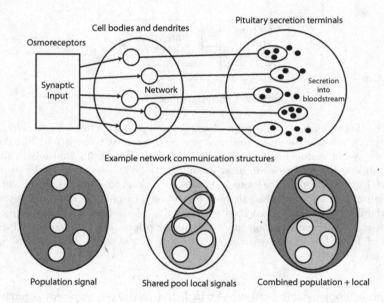

Fig. 4. The vasopressin cell population receives synaptic input from osmoreceptor neurons, and intercommunicates via dendritic release. Each cell body has its own release terminal where vesicles are released into the bloodstream. The lower panel shows varied plausible network topologies, making use of dendritic signals; a common population signal, overlapping local signal pools (similar to the oxytocin network of [29]), and the two combined.

We hypothesise that these signals coordinate the heterogeneous activity of vasopressin cells to efficiently encode osmotic stimuli over a wide dynamic range, over prolonged periods. Heterogeneity in osmosensitivity will in itself extend the dynamic range of the system, but the most active neurons will be depleted relatively rapidly. We suspect therefore that the feedback interactions may ensure that cells, during chronic stimulation, cycle through prolonged stages of activity and rest (as suggested in [19]). Such cycling may arise as an emergent property of network interactions. The network must also compensate for chronic depletion of pituitary stores and be resistant to random degradation (loss of neurons as a result, for example, of aging). We hypothesise that the network interactions will provide robustness to the system as a whole, enabling it to adapt to cell loss.

4 The Modelling Approach

Electrophysiological studies of vasopressin cells [1, 4, 34], have led to Hodgkin-Huxley type neuron models that closely match in vitro data [16, 28]. Our current single cell model incorporates the basic mechanisms implemented in these, but in a computationally simpler form – a modified leaky integrate and fire model which (with minor variations in parameters) can be tightly fit to in vivo data from the whole spectrum of recorded vasopressin cell activities, and which can therefore be duplicated with variation to form a realistic heterogeneous neuron population.

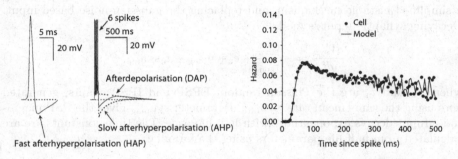

Fig. 5. Three major post spike potentials, the HAP, the AHP and the DAP shape the cells' electrical activity. The hazard, with model fitted to cell, shows how excitability changes post-spike, with shape determined by these post spike potentials. The large magnitude but fast decaying HAP generates the initial refractory period. The DAP generates the following peak in excitability which gradually falls to a plateau.

The simplest leaky integrate and fire model has a single variable and differential equation representing membrane potential; it assumes that excitability only varies with input activity, using a fixed firing threshold, and resetting after each spike. To simulate (mostly calcium driven) post-spike afterpotential dependent changes in excitability observed in vasopressin cells, our modified, non-renewal version, similar to the spike response model of Gerstner [15], adds three afterpotential variables (Fig. 5), described by ordinary differential equations as decaying exponentials, summed to generate a varied firing threshold. The transient, hyperpolarising afterpotential (HAP) causes a post spike refractory period of ~50ms. The smaller, slower, after-hyperpolarising potential (AHP) summates to limit firing rate. As the HAP decays, a subsequent depolarising afterpotential (DAP) confers a transient hyperexcitability; this can summate, contributing to the inception and maintenance of bursts. During bursts, the autocrine action of dendritic dynorphin release slowly attenuates the DAP, resulting in a shift in excitability which eventually terminates the burst. The model simulates this by adding a bistable component.

Synaptic input is modelled as a Poissonian mix of excitatory and inhibitory random perturbations to the membrane potential which leaks, decaying back towards resting potential. These perturbations are either fixed amplitude ~1-4mV or use ionic conductance based reversal potentials. Using these components, the model can match observed activity in varied vasopressin cells, matching a) spike interval distributions and hazard functions, b) burst and silence distribution, c) firing rate index of dispersion, and d) burst temporal profile. It also matches the functional behaviour of the more complex models, fit to extensive experimental data [31].

4.1 Spiking Model Equations

The current spike model is a development of the model presented in [7], using a simplified bistable mechanism, and replacing the gaussian noise based input. Decaying synaptic input is modelled by:

$$\frac{dV_{\text{syn}}}{dt} = -\frac{V_{\text{syn}}}{\tau_{\text{syn}}} + e_{\text{n}}\text{syn}_{\text{mag}} + i_{\text{n}}\text{syn}_{\text{mag}} \tag{1}$$

where e_{n} and i_{n} are the Poisson random EPSP and IPSP counts, generated here using the same mean rate, syn_{rate}. Parameter syn_{mag} gives the PSP magnitude. Input decays exponentially with half life λ_{syn}. The time constants are are calculated from half life parameters using the formula:

$$\tau_x = \frac{\lambda_x}{\log_n 2} \tag{2}$$

where x is the variable concerned.

The HAP and AHP variables H and A are modelled as decaying exponentals with halflife parameters λ_H and λ_A, incremented by k_H and k_A when a spike is fired:

$$\frac{d_H}{dt} = -\frac{H}{\tau_H} + k_H s \tag{3}$$

$$\frac{d_A}{dt} = -\frac{A}{\tau_A} + k_A s \tag{4}$$

where $s = 1$ if a spike is fired at time t, and $s = 0$ otherwise.

A similar equation describes a slow inhibitory variable I, representing dynorphin:

$$\frac{d_I}{dt} = -\frac{I}{\tau_I} + k_I s \tag{5}$$

The DAP uses a similar form, but with its value capped by parameter D_{cap}, also accounting for the effect of the AHP:

$$\frac{dD}{dt} = -\frac{D}{\tau_D} + \begin{cases} \frac{D_{cap}-D-A)}{D_{cap}}k_D s & \text{if } D + A < D_{cap} \\ 0 & \text{otherwise} \end{cases} \tag{6}$$

The following equations describe the bistable bursting mechanism:

$$\frac{dB_{syn}}{dt} = \frac{(V_{syn} - B_{syn})}{100} \tag{7}$$

$$B_i = B + B_{syn} \tag{8}$$

$$\frac{dB}{dt} = \frac{-(B_i - D + B_1 I)(B_i - I)(B_i + I)}{100} \tag{9}$$

$$V = B + H + A + V_{syn} \tag{10}$$

The bistability variable B incorporates the effects of the DAP and opposing dynorphin accumulation, encoding two stable points, bursting and silence, and an unstable balance point. Variable B_{syn} adds the random perturbations generated by synaptic input. A spike is fired when V exceeds the threshold parameter V_{th}. The resting potential is defined as 0, where the model is initialised. Parameter values were derived from fitting the model to in-vivo experimental data using a genetic algorithm, running the model on 1ms steps [7]. An example set, corresponding to Fig.1, is presented in Table 1 below.

Table 1. Fitted Model Parameters

syn_{rate}	syn_{mag}	λ_{syn}	k_H	λ_H	k_A	λ_A	k_I	λ_I	k_D	λ_D	D_{cap}	B_1	V_{th}
1240	1.2	5	-32	7.91	-0.08	1691	0.02	12081	3.4	925	12.6	0.757	12

In further work we have coupled the spiking model to a secretion model based on the physiological mechanisms of vasopressin vesicle trafficking and secretion (Fig 2). This ordinary differential equation model uses five variables; representing spike dependent changes in $[Ca^{2+}]$ concentration, activity dependent facilitation, facilitation clearance, and the size of the releasable and reserve pools. Parameters are derived from experiments, and fitted to match the non-linear stimulus-secretion coupling properties (including facilitation and fatigue) observed in vitro (e.g. [2]).

Development and testing of the model is in custom software built in C++ and wxWidgets, based on modelling and data analysis software we have previously developed to study diverse neuroendocrine systems ([24–26].

4.2   The Single Neuron Model and Communication Mechanisms

In order to simulate a heterogeneous population we must be able to duplicate the single neuron model with variation. The burst firing mechanism must be shown to remain robust under parameter variation, so that we can randomly generate a varied population, introducing variation into the model in a way that closely resembles the heterogeneity observed in-vivo.

The current stimulus-secretion components of the model represent vesicle trafficking and secretion, to reproduce facilitation and fatigue. However, testing the model over longer timescales (hours and days) must also take account of long term depletion of vasopressin stores, and will require modelling of synthesis and transport mechanisms, including long timescale store replenishment by activity-dependent synthesis [11].

The most novel part of the neuron model will be the network communication mechanisms (dendritic release and response) and tools for building the networks structure. Each released substance will have rules that govern spatial and temporal dissemination, and will have different effects. Most inhibit electrical activity, but by different mechanisms, some by modulating EPSP rate and/or IPSP rate and some post-synaptically (e.g. by modulating resting potential or EPSP magnitude). This is the stage where the project becomes more speculative and predictive. We still know very little of the functional purpose of these mechanisms, or their endogenous triggers.

4.3   Building a Robust Signal Encoding and Response Network

The simplest population signal is weak mutual interaction, which affects all neurons similarly. Dendritically-released vasopressin may be such a signal, given its abundance and long half-life. For paracrine signals (e.g. nitric oxide and endocannabinoids), or to test more limited vasopressin dispersion, each neuron will share one or more input pools (Fig 4), based on dendritic bundles [23]. We recently developed a network model to understand how oxytocin cells orchestrate synchronized bursts during reflex milk ejection [29]. In that model, each neuron contacts a few 'dendritic neighbours' via one or more shared communication pools (dendritic bundles), with dendritic secretion non-linearly coupled to firing activity; we will use similar network topologies here (Fig 4).

Choosing to model dendritic communication in this way implicitly assumes some signals are confined within defined dendritic bundles, while other signals are distributed freely throughout the population. The evidence for distinct bundles comes from electron microscopic studies showing that in dehydration, magnocellular neuron dendrites are found directly apposed in bundles of 2-8 dendrites. The bundles are enclosed by astrocyte processes that act to regulate them physiologically. Though we are not proposing to model the dynamics of bundling, this does suggest that some signals may be effectively confined within bundle based pools.

Neurons contacting two or more pools will link sub-populations, and varying the number of pools, number of neurons in each, and how many each neuron contacts, generates a wide range of structures to test. Randomly connected structures canl be built by an automated process based on defined parameters, and cell heterogeneity can be generated by randomly varying, for example, input rates and HAP parameters.

We will test network performance with increasingly difficult tasks, progressively introducing the more complex network structures. An initial network of 100 neurons will be sufficient to test varied structures but small enough to ease analysis. The first objective is a network which responds to a fixed mean input (varied between tests) linearly with a stable maintained response, proportional to the average network input over a wide dynamic range. We have good information on stimulus-secretion coupling in the nerve endings, but know less about stimulus-secretion coupling for the dendrites. However, it is activity dependent (during osmotic stimulation; [22]), so we initially assume the non-linearity of release is similar to that at the terminals, and that dendritic release is proportional to the size of the stores.

For more advanced tests, input protocols vary the input rate, either in discrete steps or continuously; the network must be able to track these changes without generating unstable feedback cycles (see Forrester flywheel) and maintain a linear output response. Neuronal responses tend to be highly non-linear, and it is still unknown what mechanism might feedback from the secretion response to modulate spike activity. We will test the network further by increasing the variation in the neurons, including excitability, bursting properties, and synthesis and secretion. The hypothesis is that burst firing is an essential element of this output response. We believe that bursting is essential for efficient signal encoding but we do not yet understand *why*.

4.4   Building and Testing a Vasopressin System

The vasopressin system is very efficient; under most conditions osmotic pressure varies only by 3%, leaving a very small dynamic range for the input signal. It can lose many of its neurons and still maintain response and can also sustain response for several days of prolonged osmotic challenge, despite limited stores. We will formulate the homeostatic control task fulfilled by the vasopressin system in a way that allows us to objectively assess its performance (and compare it with that of related networks), and to study systematically the robustness of the network in the face of a) increasing levels of noise in the inputs and b) progressive degradation, modelled as either random or activity-dependent (excitotoxic) neuronal loss.

A closed system model sufficient to relate the osmotic input signal to the secretory output will require simplified representation of water and salt intake and loss (we will not attempt to model kidney function or natriuretic mechanisms). Variables will represent total water, and salt, with differential equations to define their behaviour. Perturbations will represent drinking and eating. Salt will be lost at a concentration-dependent rate into urine. Water will be lost into urine at a rate dependent on vasopressin, and also at a constant rate, representing respiration. The network input will be a linear function of osmotic pressure (the ratio of body salt to body water), above a set point. Defining the control task in this way will enable us to study how well the system can prolong survival – i.e., when subject to chronic challenge, how long will it maintain osmolarity within tolerable bounds?

4.5 Supply and Demand

So what about the Forrester flywheel? We know that the vasopressin system is very good at this type of problem. It is able to very efficiently and robustly deliver the right amount of vasopressin, responding to highly variable demand. Our theory is that patterning in the neurons' spike response to osmostic pressure is an essential element of the system. By building a multiple spiking neuron based model, linked to secretion, we can attempt to investigate this, testing what advantage bursting might give to the dynamics of the secretion response to osmotic input. We hope to demonstate that a heterogeneous bursting population plays a role in encoding demand, and also in maintaining robust delivery while subject to varying stock levels.

5 Conclusion

This novel way of understanding the vasopressin system expresses its physiological function in terms of a defined control task, and in so doing it enables systematic, objective study of its performance, and systematic assessment of the utility of each of the various features of the vasopressin system, by evaluating the performance of closely related models in which these specific features are varied systemically. This will generate novel, testable predictions, and subsequent work will test these experimentally. Demonstrating a functional purpose for asynchronous burst firing may apply to other parts of the brain and even apply more broadly to the general problem of distributed control systems.

Acknowledgements. This project is funded by the Wellcome Trust.

References

1. Armstrong, W.E.: The Neurophysiology of Neurosecretory Cells. J. Physiol. 585, 645–647 (2007)
2. Bicknell, R.J.: Optimizing Release from Peptide Hormone Secretory Nerve Terminals. J. Exp. Biol. 139, 51–65 (1988)
3. Bie, P.: Blood Volume, Blood Pressure and Total Body Sodium: Internal Signalling and Output Control. Acta Physiol. (Oxf) 195, 187–196 (2009)
4. Bourque, C.W.: Central Mechanisms of Osmosensation and Systemic Osmoregulation. Nat. Rev. Neurosci. 9, 519–531 (2008)
5. Brown, C.H., Bourque, C.W.: Mechanisms of Rhythmogenesis: Insights from Hypothalamic Vasopressin Neurons. Trends Neurosci. 29, 108–115 (2006)

6. Buzsaki, G., Draguhn, A.: Neuronal Oscillations in Cortical Networks. Science 304, 1926–1929 (2004)
7. Clayton, T.F., Murray, A.F., Leng, G.: Modelling the In Vivo Spike Activity of Phasically-Firing Vasopressin Cells. J. Neuroendocrinol. 22, 1290–1300 (2010)
8. Del Negro, C.A., Morgado-Valle, C., Feldman, J.L.: Respiratory Rhythm: An Emergent Network Property? Neuron 34, 821–830 (2002)
9. Di, S., Boudaba, C., Popescu, I.R., Weng, F.J., Harris, C., Marcheselli, V.L., Bazan, N.G., Tasker, J.G.: Activity-dependent Release and Actions of Endocannabinoids in the Rat Hypothalamic Supraoptic Nucleus. J. Physiol. 569, 751–760 (2005)
10. Dunn, F.L., Brennan, T.J., Nelson, A.E., Robertson, G.L.: The Role of Blood Osmolality and Volume in Regulating Vasopressin Secretion in the Rat. J. Clin. Invest. 52, 3212–3219 (1973)
11. Fitzsimmons, M.D., Roberts, M.M., Robinson, A.G.: Control of Posterior Pituitary Vasopressin Content: Implications for the Regulation of the Vasopressin Gene. Endocrinology 134, 1874–1878 (1994)
12. Gouzenes, L., Desarmenien, M.G., Hussy, N., Richard, P., Moos, F.C.: Vasopressin Regularizes the Phasic Firing Pattern of Rat Hypothalamic Magnocellular Vasopressin Neurons. J. Neurosci. 18, 1879–1885 (1998)
13. Higuchi, T., Bicknell, R.J., Leng, G.:] Reduced Oxytocin Release from the Neural Lobe of Lactating Rats is Associated with Reduced Pituitary Content and Does Not Reflect Reduced Excitability of Oxytocin Neurons. J. Neuroendocrinol. 3, 297–302 (1991)
14. Hirasawa, M., Schwab, Y., Natah, S., Hillard, C.J., Mackie, K., Sharkey, K.A., Pittman, Q.J.: Dendritically Released Transmitters Cooperate via Autocrine and Retrograde Actions to Inhibit Afferent Excitation in Rat Brain. J. Physiol. 559, 611–624 (2004)
15. Jolivet, R., Lewis, T.J., Gerstner, W.: Generalized Integrate-and-fire Models of Neuronal Activity Approximate Spike Trains of a Detailed Model to a High Degree of Accuracy. J. Neurophysiol. 92, 959–976 (2004)
16. Komendantov, A.O., Trayanova, N.A., Tasker, J.G.: Somato-dendritic Mechanisms Underlying the Electrophysiological Properties of Hypothalamic Magnocellular Neuroendocrine Cells: A Multicompartmental Model Study. J. Comput. Neurosci. 23, 143–168 (2007)
17. Kondo, N., Arima, H., Banno, R., Kuwahara, S., Sato, I., Oiso, Y.: Osmoregulation of Vasopressin Release and Gene Transcription Under Acute and Chronic Hypovolemia in Rats. Am. J. Physiol. Endocrinol. Metab. 286, E337–E346 (2004)
18. Kozoriz, M.G., Kuzmiski, J.B., Hirasawa, M., Pittman, Q.J.: Galanin Modulates Neuronal and Synaptic Properties in the Rat Supraoptic Nucleus in a Use and State Dependent Manner. J. Neurophysiol. 96, 154–164 (2006)
19. Leng, G., Brown, C., Sabatier, N., Scott, V.: Population Dynamics in Vasopressin Cells. Neuroendocrinology 88, 160–172 (2008)
20. Leng, G., Ludwig, M.: Neurotransmitters and Peptides: Whispered Secrets and Public Announcements. J. Physiol. 586, 5625–5632 (2008)
21. Llorens-Cortes, C., Moos, F.: Opposite Potentiality of Hypothalamic Coexpressed Neuropeptides, Apelin and Vasopressin in Maintaining Body-Fluid Homeostasis. Prog. Brain. Res. 170, 559–570 (2008)
22. Ludwig, M., Sabatier, N., Bull, P.M., Landgraf, R., Dayanithi, G., Leng, G.: Intracellular calcium stores regulate activity-dependent neuropeptide release from dendrites. Nature 418, 85–89 (2002)
23. Ludwig, M., Leng, G.: Dendritic Peptide Release and Peptide-Dependent Behaviours. Nat. Rev. Neurosci. 7, 126–136 (2006)

24. MacGregor, D.J., Leng, G.: Modelling the Hypothalamic Control of Growth Hormone Secretion. J. Neuroendocrinol. 17, 788–803 (2005)
25. MacGregor, D.J., Lincoln, G.A.: A Physiological Model of a Circannual Oscillator. J. Biol. Rhythms 23, 252–264 (2008)
26. MacGregor, D.J., Williams, C.K., Leng, G.: A New Method of Spike Modelling and Interval Analysis. J. Neurosci. Methods 176, 45–56 (2009)
27. Ramirez, J.M., Tryba, A.K., Pena, F.: Pacemaker Neurons and Neuronal Networks: an Integrative View. Curr. Opin. Neurobiol. 14, 665–674 (2004)
28. Roper, P., Callaway, J., Armstrong, W.E.: Burst Initiation and Termination in Phasic Vasopressin Cells of the Rat Supraoptic Nucleus: A Combined Mathematical, Electrical, and Calcium Fluorescence Study. J. Neurosci. 24, 4818–4831 (2004)
29. Rossoni, E., Feng, J., Tirozzi, B., Brown, D., Leng, G., Moos, F.: Emergent Synchronous Bursting of Oxytocin Neuronal Network. PLoS Comput. Biol. 4, e1000123 (2008)
30. Ruan, M., Brown, C.H.: Feedback Inhibition of Action Potential Discharge by Endogenous Adenosine Enhancement of the Medium Afterhyperpolarization. J. Physiol. 587, 1043–1066 (2009)
31. Sabatier, N., Brown, C.H., Ludwig, M., Leng, G.: Phasic Spike Patterning in Rat Supraoptic Neurones In Vivo and In Vitro. J. Physiol. 558, 161–180 (2004)
32. Sabatier, N., Leng, G.: Bistability with Hysteresis in the Activity of Vasopressin Cells. J. Neuroendocrinol. 19, 95–101 (2007)
33. Stern, J.E., Zhang, W.: Cellular Sources, Targets and Actions of Constitutive Nitric Oxide in the Magnocellular Neurosecretory System of the Rat. J. Physiol. 562, 725–744 (2005)
34. Tasker, J.G., Di, S., Boudaba, C.: Functional Synaptic Plasticity in Hypothalamic Magnocellular Neurons. Prog. Brain. Res. 139, 113–119 (2002)
35. Towill, D.R.: Industrial Dynamics of Modelling Supply Chains. International Journal of Physical Distribution & Logistics 26, 23–42 (1996)

The Effective Calcium/Calmodulin Concentration Determines the Sensitivity of CaMKII to the Frequency of Calcium Oscillations

Thiago M. Pinto, Maria J. Schilstra, and Volker Steuber

Science and Technology Research Institute,
University of Hertfordshire, Hatfield, Herts, AL10 9AB, UK
{t.pinto,m.j.1.schilstra,v.steuber}@herts.ac.uk

Abstract. Calcium/calmodulin-dependent protein kinase II (CaMKII) is involved in the induction of many forms of synaptic plasticity in the brain. Experimental and computational studies have shown that CaMKII is sensitive to the frequency of oscillatory Ca^{2+} signals. Here we demonstrate that in a simple, commonly used kinetic model of CaMKII phosphorylation, the overall phosphorylation rate under sustained application of $Ca_4 - CaM$ pulses ultimately depends on the average ('effective') concentration of $Ca_4 - CaM$ in the system, rather than on the pulse frequency itself. As a corollary, equal phosphorylation levels are achieved in response to pulsed and constant applications of equal effective concentrations of $Ca_4 - CaM$.

Keywords: CaMKII, calmodulin, Ca^{2+} oscillations.

1 Introduction

Calcium/calmodulin-dependent protein kinase II (CaMKII), which is present in high concentrations in the brain, is a multifunctional protein kinase involved in Ca^{2+} signalling systems that underlie the induction of synaptic plasticity. Brief Ca^{2+} signals can activate CaMKII, and stimulate an autophosphorylation reaction that allows the kinase to maintain its activation level [1].

Earlier computer simulations based on a widely used CaMKII autophosphorylation model indicated that CaMKII activation is sensitive to the frequency of Ca^{2+} oscillations [2–5], and *in vitro* experiments have demonstrated that the kinase does indeed respond differently to different frequencies of Ca^{2+} spikes [6].

Here, we present a somewhat simplified version of the CaMKII activation model developed by Dupont et al. [5], and show that this model reproduces the results of the more complex one. Further, we demonstrate that CaMKII activation by Ca^{2+} and calmodulin (CaM) in the model is mostly determined by the effective $Ca_4 - CaM$ concentration, which varies with the frequency of $Ca_4 - CaM$ pulses, but does not depend on the actual frequency of Ca^{2+} oscillations. Moreover, we show that the application of a constant level of $Ca_4 - CaM$ with the same mean concentration as in the pulsed protocol results in the same level of CaMKII phosphorylation.

M.A. Lones et al. (Eds.): IPCAT 2012, LNCS 7223, pp. 131–135, 2012.
© Springer-Verlag Berlin Heidelberg 2012

2   CaMKII Activation Model

In the simplified CaMKII activation model (Fig. 1), CaMKII can be in four different states: inactive (W_i), bound to $Ca_4 - CaM$ (W_b), phosphorylated and bound to $Ca_4 - CaM$ (W_p), and autonomous (W_a): phosphorylated, but dissociated from $Ca_4 - CaM$. As in the earlier model [5], we assume that $Ca_4 - CaM$ formation is rapid and complete. Binding of $Ca_4 - CaM$ to a CaMKII subunit results in the activation of the subunit's kinase function, allowing it to phosphorylate its substrates. These substrates include the subunit's nearest neighbours in the CaMKII multimer. The autophosphorylation rate associated with this process is indicated as V_a, and is described using the phenomenological nonlinear function of the concentrations of W_b, W_p and W_a as in [5]. Dissociation of $Ca_4 - CaM$ from the phosphorylated form yields the so-called autonomous form of CaMKII, which retains some or all of its kinase activity.

Different from the Dupont model [5], we do not model a "trapped" state [7], mainly because dissociation of Ca^{2+} and CaM cannot be distinguished experimentally (nor described thermodynamically) as two distinct processes.

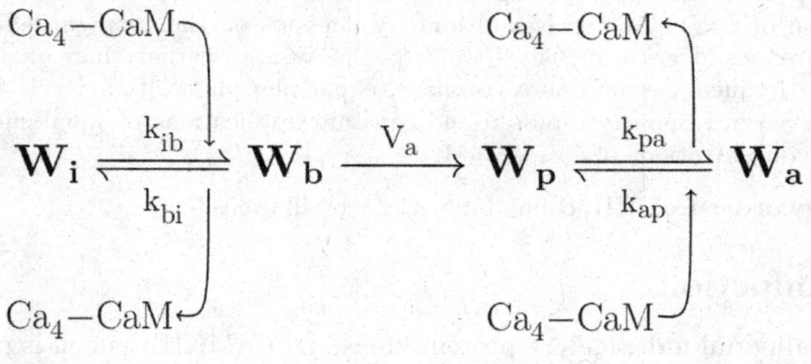

Fig. 1. Model of the activation of CaMKII by $Ca_4 - CaM$. k_{ib}, k_{bi}, k_{pa} and k_{ap} are the rate constants of the reversible $Ca_4 - CaM$ binding reactions, and V_a is the rate of the irreversible phosphorylation of W_b

The model was implemented as a set of 4 coupled ordinary differential equations (ODEs), which were solved numerically using the XPPAUT software (X-Windows Phase Plane plus Auto).

3   Results

To examine whether the omission of trapped state had any significant effect, and to investigate the dependence of the overall autophosphorylation kinetics on the frequency of $Ca_4 - CaM$ oscillations, we replicated the simulations presented in [5] with our simplified model. Figure 2b shows the simulated response to 100

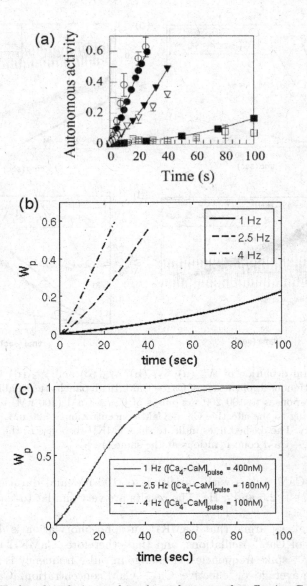

Fig. 2. CaMKII phosphorylation and its dependence on the effective $Ca_4 - CaM$ concentration. **(a)** Temporal evolution of the phosphorylated form of CaMKII (W_p) in response to 100 200 ms square pulses of $Ca_4 - CaM$ (100 nM) at frequencies of 1 Hz (solid squares), 2.5 Hz (solid triangles) and 4 Hz (solid circles) in the Dupont model [5]. **(b)** W_p in response to the same protocol in our simplified model. **(c)** W_p in response to 100 200 ms square pulses of $Ca_4 - CaM$ at 1, 2.5 and 4 Hz, but with scaled pulse amplitudes so that the effective concentration of $Ca_4 - CaM$ is 80 nM. The amplitudes of $Ca_4 - CaM$ pulses are 400 nM at 1 Hz (solid), 160 nM at 2.5 Hz (dashed) and 100 nM at 4 Hz (dashed-dotted).

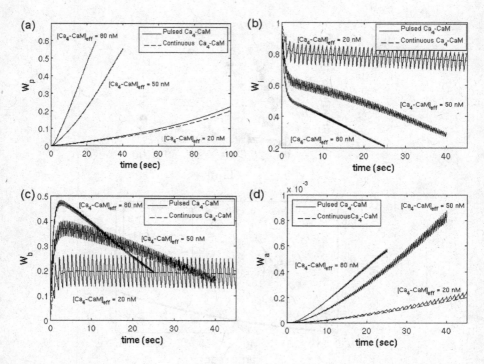

Fig. 3. Temporal evolution of W_p **(a)**, W_i **(b)**, W_b **(c)** and W_a **(d)** in response to pulsed and continuous applications of $Ca_4 - CaM$. In all panels, the solid lines represent the CaMKII responses to 100 200 ms pulses of $Ca_4 - CaM$ (100 nM) at 1, 2.5 and 4 Hz, corresponding to the effective $Ca_4 - CaM$ concentrations of 20 nM, 50 nM and 80 nM, respectively. The dashed lines indicate the CaMKII response to the application of continuous $Ca_4 - CaM$ concentrations at the same levels.

square $Ca_4 - CaM$ pulses with amplitude of 100 nM and duration of 200 ms each, applied at 1, 2.5 and 4 Hz. The results are very similar to those presented in [5] (cf Fig. 2a).

Dupont et al. [5] argue that CaMKII autophosphorylation is dependent on the frequency of Ca^{2+} oscillations, and that, therefore, CaMKII may act as a decoder of Ca^{2+} spike frequencies. A change in pulse frequency is accompanied by a change in average, or 'effective' $Ca_4 - CaM$ concentration, $[Ca_4 - CaM]_{eff}$, which is computed as $[Ca_4 - CaM]_{eff} = [Ca_4 - CaM]_{pulse} \times f \times L$ (where $[Ca_4 - CaM]_{pulse}$ is the pulse amplitude, f is the pulse frequency, and L the pulse duration). Thus, CaMKII is subjected to different effective $Ca_4 - CaM$ concentrations, which in turn affects the average concentration of W_b and W_p, and the autophosphorylation kinetics. To investigate whether the autophosphorylation kinetics are primarily determined by the pulse frequency, or by the accompanying variation in $[Ca_4 - CaM]_{eff}$, we rescaled the $Ca_4 - CaM$ concentrations to an equal effective concentration of 80 nM, and compared the phosphorylation kinetics.

Figure 2c shows that the phosphorylation kinetics at 1, 2.5 and 4 Hz pulses are identical after rescaling the $Ca_4 - CaM$ concentration. This strongly indicates that, at least for the frequency range examined here, the CaMKII autophosphorylation kinetics are independent of the pulse frequency itself.

We also investigated whether the autophosphorylation kinetics are the same under constant and pulsed $Ca_4 - CaM$ concentrations at the appropriate effective concentrations. Figure 3 shows a superposition of concentrations of the various species under pulsed and continuous $Ca_4 - CaM$ concentration conditions. Again, these results indicate that the CaMKII autophosphorylation kinetics in this model are determined by the effective $Ca_4 - CaM$ concentration, not by the actual pulse frequency.

References

1. Hanson, P., Schulman, H.: Neuronal Ca2+/calmodulin-dependent protein kinases. Annual Review of Biochemistry 61(1), 559–601 (1992)
2. Hanson, P., Meyer, T., Stryer, L., Schulman, H.: Dual role of calmodulin in autophosphorylation of multifunctional CaM kinase may underlie decoding of calcium signals. Neuron 12(5), 943–956 (1994)
3. Michelson, S., Schulman, H.: CaM kinase: A model for its activation and dynamics. Journal of Theoretical Biology 171, 281–290 (1994)
4. Dosemeci, A., Albers, R.: A mechanism for synaptic frequency detection through autophosphorylation of CaM kinase II. Biophysical Journal 70(6), 2493–2501 (1996)
5. Dupont, G., Houart, G., De Koninck, P.: Sensitivity of CaM kinase II to the frequency of Ca2+ oscillations: a simple model. Cell Calcium 34(6), 485–497 (2003)
6. De Koninck, P., Schulman, H.: Sensitivity of CaM kinase II to the frequency of Ca2+ oscillations. Science 279(5348), 227–230 (1998)
7. Waxham, M., Putkey, J., Tsai, A.: A mechanism for calmodulin (CaM) trapping by CaM-kinase II defined by a family of CaM-binding peptides. Journal of Biological Chemistry 273(28), 17579–17584 (1998)

The Effect of Different Types of Synaptic Plasticity on the Performance of Associative Memory Networks with Excitatory and Inhibitory Sub-populations

Alex Metaxas, Reinoud Maex, Volker Steuber,
Rod Adams, and Neil Davey

Science and Technology Research Institute,
University of Hertfordshire, UK
{a.metaxas,r.maex1,v.steuber,r.g.adams,n.davey}@herts.ac.uk

Abstract. In real neuronal networks it is known that neurons are either excitatory or inhibitory. However, it is not known whether all synapses within the subpopulations are plastic. It is interesting to investigate the implications these constraints may have on functionality. Here we investigate highly simplified models of associative memory with a variety of allowed synaptic plasticity regimes. We show that the allowed synaptic plasticity does indeed have a large effect on the performance of the network and that some regimes are much better than others.

Keywords: Synaptic plasticity, associative memory, sign constraint, excitatory/inhibitory neurons.

1 Introduction

Synaptic plasticity has long been implicated in the storage of memories in the brain. While there has been much research into plasticity in excitatory neurons, it is only recently that the range of types and possible functions of inhibitory plasticity have been investigated. The view that inhibitory interneurons provide a purely regulatory role is being challenged as diverse forms of interneuron plasticity are discovered [1].

Here we present early work investigating the role of varying types of synaptic plasticity in memory storage and retrieval. In an initial set of simulations we use associative memory networks of perceptrons. These are trained using a modified version of the perceptron learning rule and we subsequently measure their performance by determining their ability to store and recall patterns [2]. The networks consist of subpopulations of excitatory and inhibitory units. The role of plasticity at inhibitory synapses in these types of networks has only recently begun to be studied systematically.

While we find that sufficient synaptic plasticity is required for the network to function effectively, a lack of plasticity can be tolerated in some types of synaptic connection. The size of the weight of the non-plastic connections is also important in determining the characteristics of the function of the network.

M.A. Lónes et al. (Eds.): IPCAT 2012, LNCS 7223, pp. 136–142, 2012.
© Springer-Verlag Berlin Heidelberg 2012

2   Materials and Methods

2.1   Network Models

We use networks of fully connected perceptrons, divided into excitatory and inhibitory sub-populations, meaning their efferent synapses are non-negative and non-positive respectively. In the mammalian cortex it is thought that there is a ratio of 80:20 excitatory to inhibitory neurons [3]. We therefore define the population bias as the ratio of inhibitory to excitatory neurons in the network. This is set to either 0.2 or 0.5, where 0.2 implies 20% (80%) inhibitory (excitatory) neurons and 0.5 implies a balanced distribution. The networks function as Hopfield-like associative memories, storing binary patterns in the synaptic weights. The units are updated asynchronously, in fixed order and when presented with corrupted or noisy patterns the networks will dynamically change to the original stored pattern. In this work we consider both the inhibitory and excitatory neurons to form the patterns.

It is known that in real neuronal systems coding is often sparse. We therefore use two types of pattern encodings. Firstly, where the number of on bits and off bits in the pattern are balanced, giving a pattern bias of 1s to 0s of 0.5, and secondly pattern bias of 0.2 that corresponds to sparser patterns (ones with fewer on bits). The patterns are randomly generated and uncorrelated.

As mentioned above, each unit of the network is a perceptron. Their behavior is governed by the following equation:

$$y = \begin{cases} 1 \; if \; \boldsymbol{w} \cdot \boldsymbol{x} > 0 \\ 0 \; otherwise \end{cases} \tag{1}$$

where $\boldsymbol{w}$ and $\boldsymbol{x}$ are vectors of real-valued weights and binary states of the afferent presynaptic perceptrons and 0 is the activation threshold. The scalar output y is binary.

2.2   Plasticity

In the networks there are four types of synaptic connections determined by the nature of the pre- and post- synaptic perceptrons: Inhibitory-to-Inhibitory (I-to-I), Excitatory-to-Excitatory (E-to-E), Inhibitory-to-Excitatory (I-to-E) and Excitatory-to-Inhibitory (E-to-I). Starting with the assumption that E-to-E connections are known to be plastic, we rule out half of the possible 16 configurations. We then select the configuration where inhibitory synapses are non-plastic (Fixed I-to-*) as was previously thought [4], and two variations (Fixed I-to-I, Fixed *-to-I). Finally we select the extreme cases where all synapses are plastic (Free) or only E-to-E are plastic (Free E-to-E). Each plasticity configuration (see Table 1) describes which synaptic connections may be plastic and which are set to a fixed weight. In the first experiment the fixed weights are set to either 1 or N, so that they either vary or do not vary with the size N of the network. In the second experiment we fix the values of all other parameters and vary the fixed weight over a range of values to explore its effect in more detail.

Table 1. Plasticity Configuration. ✓ 's denote the connection is plastic; ✗'s denote the connection is fixed.

	Free	Free E-to-E	Fixed I-to-I	Fixed I-to-*	Fixed *-to-I
I-to-I	✓	✗	✗	✗	✗
I-to-E	✓	✗	✓	✗	✓
E-to-E	✓	✓	✓	✓	✓
E-to-I	✓	✗	✓	✓	✗

Note that the connections defined as plastic are subject to change only during the learning phase (described below) and do not change during the recall phase.

2.3   Learning

Canonical associative memory models with threshold units, for example the Hopfield net, commonly use a one-shot Hebbian learning rule. However, this form of learning does not perform well when the number of patterns is large and in this work we use a modified version of the standard perceptron learning rule [5]. We use training sets of random binary vectors with varying bias (see above).

We also introduce some modifications to the learning rule. The first modification ensures that the efferent (outgoing) weights of units that we have specified as excitatory or inhibitory maintain the correct sign constraint. Namely, an excitatory (inhibitory) perceptron has only non-negative (non-positive) outgoing weights. We do this by preventing excitatory (inhibitory) weights from going below (above) zero – should they wish to do so they are thresholded at zero. The second modification introduces a check to determine if the connection between the pre- and post-synaptic units has been marked as fixed or plastic. If fixed, the weight is set at the specified fixed weight size and does not change.

2.4   Measures of Memory Performance

The associative memory performance of the threshold unit network is measured by its *Effective Capacity* (*EC*) [6]. Effective Capacity is a measure of the maximum number of patterns that can be stored in the network with reasonable pattern correction still taking place. In other words, it is a capacity measure that takes into account the dynamic ability of the network to perform pattern correction. We take a fairly arbitrary definition of *reasonable* as the ability to correct the addition of 60% noise (selecting 60% of the bits and then flipping them with 0.5 probability, resulting in ~30% pattern corruption) to within a similarity of 95% with the original fundamental memory, that is, 95% of the bits are identical. Varying these two percentage figures gives differing values for *EC* but the values with these settings are robust for comparison purposes. For large fully connected networks the *EC* value is about 0.1 of the conventional capacity of the network, but for networks with sparse, structured connectivity *EC* is dependent upon the actual connectivity pattern.

The Effective Capacity of the network is therefore the highest pattern loading for which all 60% corrupted versions of the stored patterns have, after convergence, a mean similarity of 95% or greater with their original state.

2.5 Simulator

All experiments were run on our internally developed neural network simulator. The simulator currently consists of a C/C++ library that is GPU accelerated via CUDA C/C++. This simulator is still at an early stage of development and it is our intention to provide a Python interface (to facilitate scripted experiments) and to support spiking neural networks in the future.

3 Results

3.1 Varying Plasticity

With the five plasticity configurations we've already identified we also vary three parameters, the population bias, pattern bias, and the fixed weight (as described above), between two levels, in networks of $N = 1000$ units and measure their performance by effective capacity (see above).

Figure 1 shows that the optimal configuration for the Free (or fully plastic) network is to have both balanced population and pattern biases (A & B). We can see that the effect of reducing (to 0.2) either the population (E & F) or pattern (C & D) biases is roughly the same in the Free networks. The effect of reducing them both to 0.2 results in a further reduction in performance. In a comparable network in which the units are not sign constrained and synapses can take any value, the measured EC is about 200. All of the Free networks perform reasonably well and the optimal cases (A & B) show results consistent with the prediction that sign constrained networks have half the actual capacity of unconstrained networks [7].

The Free E-to-E (in which, in the worst case, only 25% of all the synapses in the network are plastic) and Fixed *-to-I (50% plastic in the worst case) configurations have too little plasticity, resulting in an EC of 0. The reason for this is that only 15-50% (depending on the parameter set) of the units are able to converge during training (within 10000 epochs) in both the case of the Free E-to-E and Fixed *-to-I, preventing the initial patterns from being stored correctly as attractors.

While there is a large difference in the performance of Fixed I-to-I and Fixed I-to-*, it is clear that the size of the fixed weight is important. In almost all cases, if the fixed weight is set to N the network has a significant reduction in effective capacity.

We see that in the case of Fixed I-to-I (4% when the population bias is 0.2), a configuration with only limited fixed connections, reasonable performance can be maintained (A & D). The higher EC of parameter set D can be explained by the smaller inhibitory population size and hence the reduced number of fixed connections. In fact in network A, only 75% of the synapses are plastic.

The key results are that these networks can tolerate a limited level of fixed connections, and that the size of the fixed weight could be important.

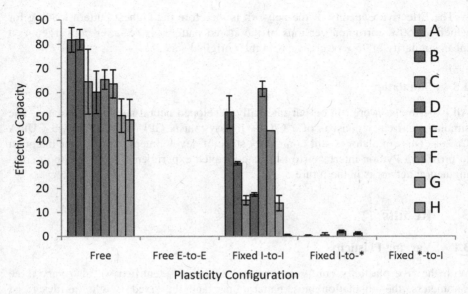

Fig. 1. Memory performance for varying plasticity rules within and between the excitatory and inhibitory subpopulations (see Table 1 for rule description). The eight colours indicate different parameter sets for the three studied parameters. (**A**) Population bias: 0.5, pattern bias: 0.5, fixed weight size 1. (**B**) Population bias: 0.5, pattern bias: 0.5, fixed weight size N. (**C**) Population bias: 0.5, pattern bias: 0.2, fixed weight size 1. (**D**) Population bias: 0.5, pattern bias: 0.2, fixed weight size N. (**E**) Population bias: 0.2, pattern bias: 0.5, fixed weight size 1. (**F**) Population bias: 0.2, pattern bias: 0.5, fixed weight size N. (**G**) Population bias: 0.2, pattern bias: 0.2, fixed weight size 1. (**H**) Population bias: 0.2, pattern bias: 0.2, fixed weight size N. Error bars show one standard deviation.

3.2   Effect of Fixed Weight Size on Performance

We vary the magnitude of the fixed weight in networks of $N = 100$ units in a Fixed I-to-I configuration. The population and pattern biases are 0.5.

The optimal size of the fixed weight in our experiments appears to be around 1. In networks with this value, the pre-synaptic weights of the inhibitory units were generally of the same order of magnitude as those of the excitatory units. We hypothesized this might be the case because the mean size of the weights in the plastic synapses was around one.

In order to check this hypothesis we calculated the mean of the weights in the Free network, and its order of magnitude was indeed one (e.g. -0.59/0.54 for inhibitory and excitatory weights respectively).

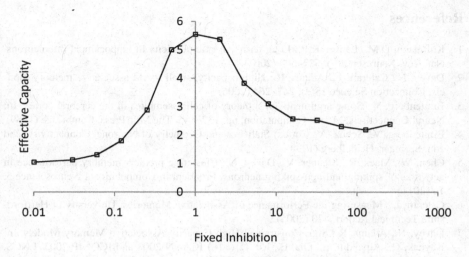

Fig. 2. Memory performance of Fixed I-to-I networks with varying fixed weights. Note the logarithmic scale.

4   Discussion

Networks with subpopulations of excitatory and inhibitory units can perform well as associative memory. In our best performing network in which the proportion of excitatory and inhibitory units was the same, the effective capacity is just under half that of a completely unconstrained network without separate subpopulations of excitatory and inhibitory neurons. Moreover, under some circumstances a proportion of the units can have fixed non-learnable weights, with the network still functioning acceptably. In particular, networks in which all the inhibitory to inhibitory weights are fixed, in one case a quarter of the weights in the whole network, still perform reasonably well. However, it appears that several other restrictions on learnability have severe effects on the ability of the network to function. Unsurprisingly, if too many weights are not plastic then the network will fail to learn properly.

The next question was whether the specific value of the fixed weight was significant. Counter intuitively, it appears that the actual value is important in determining how the network performs. This seems to be the case because, in these networks, the mean weight size of the trainable weights, post training, is of order of magnitude one. It is worth noting that the fixed weight does not appear to need to be changed with the network size (as described above).

We are currently extending our results by investigating networks of spiking neurons, other connectivity patterns, and where only excitatory units are considered part of the pattern.

References

1. Kullmann, D.M., Lamsa, K.P.: Long-term synaptic plasticity in hippocampal interneurons. Nat. Rev. Neurosci. 8(9), 687–699 (2007)
2. Davey, N., Calcraft, L., Adams, R.: High capacity, small world associative memory models. Connection Science 18(3), 247–264 (2006)
3. Braitenberg, V.: Some arguments for a theory of cell assemblies in the cerebral cortex. In: Neural Connections, Mental Computation, pp. 137–145. The MIT Press, Cambridge (1989)
4. Braitenberg, V., Schuz, A.: Cortex: Statistics and Geometry of Neuronal Connectivity, 2nd edn. Springer, Heidelberg (1998)
5. Chen, W., Maex, R., Steuber, V., Davey, N.: Clustering predicts memory performance in networks of spiking and non-spiking neurons. Frontiers in Computational Neuroscience 5, 14 (2011)
6. Calcraft, L.: Measuring the Performance of Associative Memories. University of Hertfordshire Technical Report, 420 (2005)
7. Davey, N., Adams, R.: Sign Constrained High Capacity Associative Memory Models. In: Kaynak, O., Alpaydın, E., Oja, E., Xu, L. (eds.) ICANN 2003 and ICONIP 2003. LNCS, vol. 2714, pp. 78–81. Springer, Heidelberg (2003)

Simulating Neurons
in Reaction-Diffusion Chemistry

James Stovold and Simon O'Keefe

Department of Computer Science, University of York

Abstract. Diffusive Computation is a method of using diffusing particles as a representation of data. The work presented attempts to show that through simulating spiking neurons, diffusive computation has at least the same computational power as spiking neural networks. We demonstrate (by simulation) that wavefronts in a Reaction-Diffusion system have a cumulative effect on concentration of reaction components when they arrive at the same point in the reactor, and that a catalyst-free region acts as a threshold on the initiation of an outgoing wave. Spiking neuron models can be mapped onto this system, and therefore RD systems can be used for computation using the same models as are applied to spiking neurons.

Diffusive Computation is a method of using diffusing particles as a representation of data. The diffusion of these particles can be used for various forms of information processing. The use of a diffusive chemical reaction has been studied in great depth since it was proposed as a basis for morphogenesis by Turing [15].

In [6], the generality of the classical computing paradigm is analysed in comparison to how exploitable the complexity of alternative forms of computing are. One such alternative is diffusive computation. The work presented herein attempts to show that through simulating spiking neurons, diffusive computation has at least the same computational power as spiking neural networks.

Reaction-Diffusion Chemistry (RD Chemistry) is a chemical reaction that changes state in such a way that wavefronts of reagent appear to flow (or diffuse) across the solution. The most widely-used model is based on the ferroin-catalysed Belousov-Zhabotinsky (BZ) reaction [5], which may display single waves of reaction or an oscillation between two (observable) states. The BZ reaction is often modelled numerically using the Oregonator model [9], which consists of three second-order partial differential equations (PDEs) that describe how the concentration of each reagent changes over time. The BZ reaction was used as the basis for the simpler cerium-catalysed Rovinsky-Zhabotinsky (RZ) model [13], which is defined mathematically as two second-order PDEs (given in Appendix A). The results presented are based on the RZ model of a BZ reaction.

The role of the catalyst in the reaction is central to the way in which the RD system can be controlled and therefore viewed as a computing system. In a thin-layer reactor containing a BZ system in solution, if the reaction is initiated at a particular location then waves will spread out from that point, showing the change in state of the reaction by a change in the oxidation state (and therefore

M.A. Lones et al. (Eds.): IPCAT 2012, LNCS 7223, pp. 143–149, 2012.

colour) of the catalyst. If the further step is taken of binding the catalyst to a substrate, and restricting it to be present only in certain regions or channels, then the reaction can only take place in these regions and we observe wavefronts that appear to be constrained to particular channels. The important point is that diffusion of the non-catalyst chemistry may take place, and if there is a sufficient concentration then it may diffuse across a non-catalytic gap and start a catalysed reaction on the far side [11].

RD chemistry can be used experimentally as a form of diffusive computing, and has been investigated for various different applications [2,7,10,14,15]. In [7], it is suggested that a 'train' of diffusing chemical waves can overcome a larger gap in catalytic material than a single chemical wave. This can be interpreted as a spike train in a spiking neural network, where the neuronal membrane potential is modelled by the concentration of activator reagent in each neuron. The release of activator into the non-catalytic gap by incoming waves is analogous to the process of neurotransmitter release into the synaptic cleft and subsequent ion channel activation — the concentration of reagent in the gap raises the concentration of the post-synaptic neuron through a very similar diffusive method as in real neurons. The motivation here is to demonstrate sufficient similarity between this view of an RD system and models of spiking neurons to allow the transfer of models of computation from the cell to the RD system.

In considering computation in real (and therefore spiking) neurons, we usually consider an abstract or artificial neuron model. In its simplest formulation, an artificial neuron receives inputs that are binary (0,1) and produces an output if the sum of the inputs exceeds a threshold, i.e.

$$y = \begin{cases} 1 : \sum_i x_i > \theta \\ 0 : \sum_i x_i \leq \theta \end{cases}$$

Such binary neurons can compute functions such as element distinctness using only a single neuron, whereas typically this requires $O(log(n!))$ binary threshold gates [16]. Therefore demonstrating a single spiking neuron simulated in RD chemistry will show that we can do useful computation with this view of the RD system.

The work presented herein suggests that by organising a chemical circuit such that multiple waves arrive at the 'synapse' in close succession, the 'neuronal membrane potential' in the region beyond the synapse rises high enough to form a new spike; Fig. 1 shows a simple simulation of an RD system demonstrating this (details of the simulation are given in Appendix A). The first two frames show the diffusing waves before they arrive at the non-catalytic gap (dark areas) and the third frame shows the resultant wave in the region beyond the synapse.

There is a significant time lag between the arrival of waves at the synaptic cleft and the formation of a new wave on the post-synaptic side. This is similar to the delay exhibited by chemical synapses between neurotransmitter release and the post-synaptic action potential. The ability to control the delay is essential to

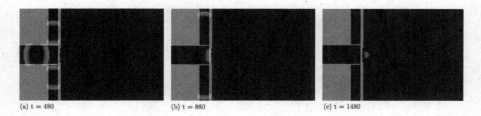

(a) t = 480   (b) t = 880   (c) t = 1480

Fig. 1. Simulation of a chemical synapse in a reaction-diffusion medium: (a) wave approaching synaptic cleft, (b) wave enters synapse, (c) post-synaptic response

be able to map computational models that include explicit delays from spiking neurons into the RD domain.

Fig. 2 shows a trace of a single pixel in simulation (at location $x = 42$, $y = 50$). The first trace is the concentration of activator reagent when a single wave attempts to cross the synaptic cleft. In this case, the concentration remains below the threshold and no post-synaptic spike is produced. The second is the concentration when multiple waves (in this case, three) attempt to cross the cleft. Due to the higher concentration of activator reagent present in the cleft after multiple waves, the concentration of the post-synaptic neuronal membrane potential is high enough (at time $t \approx 1200$) to cross the threshold required to 'fire' an action potential ($t \approx 1500$).

This experiment is a first step toward concluding that diffusive media can be used to construct a neural network. This would have significant implications of the computational power present in diffusive media, in particular exploiting the inherent parallelisability of neural networks.

The work presented above allows for spatial correlation of spikes (i.e. multiple spikes arriving from different sources), but to build a full network, it is also necessary to be able to correlate spikes temporally. The current progress of this work is shown in Fig. 3, where the two panes show the current architecture for spatial correlation and proposed architecture for temporal correlation respectively.

The architecture for temporal correlation is designed to take into account the difficulty of maintaining two waves very close together — the refractory period from the first wave stops the second wave if it follows too close behind. So, as shown in Fig. 3, two 'delay loops' are included. This allows the first wave to 'load' the delay loops, but will not pass across the synapse. If the second wave arrives at the correct time after the first, it will arrive at the synapse at the same time as the delay loops unload. As in the spatially-correlated version, the three waves presented to the synapse should then allow the wave to cross the synaptic cleft. The length of the path in the delay loops dictates how closely the second wave must follow the first.

Once this has been completed, the next step is to mathematically identify the relationship between the neuronal threshold, θ, and the size of non-catalytic gap,

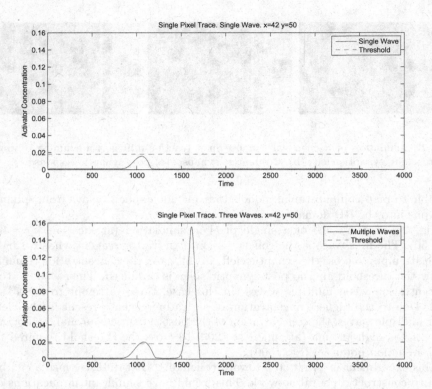

Fig. 2. Graph showing the activator response (at location $x = 42$, $y = 50$) to single and multiple chemical waves: (top) single wave, doesn't reach threshold; (bottom) three waves, reaches threshold and fires action potential

in particular looking at the potential of modelling ligand-gated ion channels as a way of varying the post-synaptic neuronal threshold (essentially the addition of protruding lines of catalytic material into the synaptic cleft to increase the excitability of the post-synaptic side).

The work described above uses controlled distribution of a catalyst to implement what is effectively a 2D architecture for the propagating waves. Given the right chemical conditions, an RD system may also exhibit the propagation of wavelets — small fragments of a wavefront that propagate in a particular direction, remaining about the same size, without the need to constrain the distribution of catalyst [3]. Interaction between these wavelets is an alternative way of implementing a spiking neuron. However, in a 'large' reactor (relative to the sub-millimetre scale of the wavelets), designing and sustaining the trajectories of a large number of wavelets would be difficult.

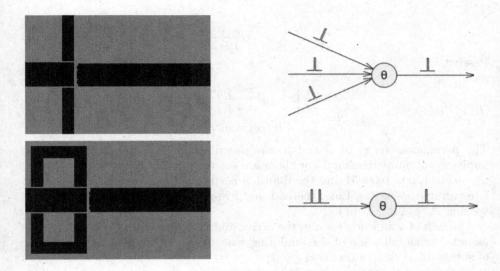

Fig. 3. (top) Current architecture for spatially-correlated spike trains; (bottom) Proposed architecture for temporally-correlated spike trains

However, [1] proposes an interesting use of a lipid membrane to form vesicles of reagent, which can be used to constrain wave propagation in RD systems to small regions, with communication between neighbouring vesicles. The added constraints on wave propagation make the routing of wavelets across the reactor easier. This suggests a mechanism for 'networking' a number of simple neurons that operate as described above.

After constructing a full spiking neural network, an appropriate move would be to look into modelling associative memory style networks such as Correlation Matrix Memories (CMMs [4]). The ease of training CMMs compared with other neural networks make them an ideal move for consolidating the computational power of Diffusive Computing.

Appendix A: Simulation Details

As discussed above, the simulator is based on the Rovinsky-Zhabotinsky model [13], a simplified version of the Belousov-Zhabotinsky reaction [5]. Modelling the RZ system numerically is described by second-order partial differential equations over two variables, x (activator) and z (inhibitor).

The simulator distinguishes between active (catalytic) regions and passive (non-catalytic) regions as in [8]:

Active:
$$\frac{\partial x}{\partial t} = \frac{1}{\epsilon}\left[x(1-x) - \left(2q\alpha\frac{z}{1-z} + \beta\right)\frac{x-\mu}{x+\mu}\right] + D_x\nabla^2 x$$

$$\frac{\partial z}{\partial t} = x - \alpha \frac{z}{1 - z}$$

Passive:

$$\frac{\partial x}{\partial t} = -\frac{1}{\epsilon}\left[x^2 + \beta\frac{x - \mu}{x + \mu}\right] + D_x \nabla^2 x$$

$$z = 0 \text{ (constant)}$$

The parameters (ϵ, q, α, β and μ) are described in [12], and in the current implementation of the simulator these are set to: $\epsilon = 2.5$, $q = 0.5$, $\alpha = 0.068$, $\beta = 0.0034$, $\mu = 0.00051$ and the diffusion coefficient, $D_x = 0.002$. The system is integrated using the Euler method and five-point Laplacian with grid-point spacing $\Delta x = 0.15$, as in [2].

The lack of a diffusion term in the active inhibitor equation ($\frac{\partial z}{\partial t}$) is due to the assumed immobilisation of the inhibiting reagent, i.e. by binding to some form of substrate, such as a chemical gel [2].

The state of the reaction is stored in two variables, the activator and inhibitor concentrations and a constant flag to designate the presence of catalyst. Initially, the catalyst is mapped out and all areas of space are set to the neural concentration (0.0) for both activator and inhibitor. The waves start from a point of stimulation where the activator concentration of 10 adjacent pixels are set to 1.0.

References

1. Adamatzky, A., Bull, L., De Lacy Costello, B., Holley, J., Jahan, I.: Computational Modalities of Belousov-Zhabotinsky Encapsulated Vesicles. ArXiv e-prints (September 2010)
2. Adamatzky, A.: Collision-based computing in Belousov-Zhabotinsky medium. Chaos, Solitons & Fractals 21(5), 1259–1264 (2004)
3. Adamatzky, A., De Lacy Costello, B.: Binary collisions between wave-fragments in a sub-excitable Belousov-Zhabotinsky medium. Chaos, Solitons & Fractals 34(2), 307–315 (2007)
4. Austin, J., Stonham, T.: Distributed associative memory for use in scene analysis. Image and Vision Computing 5(4), 251–260 (1987)
5. Belousov, B.P.: A periodic reaction and its mechanism. Med. Publ., Moscow (1959)
6. Conrad, M., Zauner, K.: Molecular Computing Conformation-Based Computing: A Rationale and a Recipe, pp. 1–31. MIT Press (2003)
7. Gorecki, J., Gorecka, J., Igarashi, Y.: Information processing with structured excitable medium. Natural Computing 8, 473–492 (2009)
8. Gorecki, J., Yoshikawa, K., Igarashi, Y.: On chemical reactors that can count. The Journal of Physical Chemistry A 107(10), 1664–1669 (2003)
9. Janz, R.D., Vanecek, D.J., Field, R.J.: Composite double oscillation in a modified version of the oregonator model of the Belousov-Zhabotinsky reaction. The Journal of Chemical Physics 73(7), 3132–3138 (1980)
10. Kuhnert, L., Agladze, K.I., Krinsky, V.I.: Image processing using light-sensitive chemical waves. Nature 337(6204), 244–247 (1989)

11. Motoike, I., Yoshikawa, K.: Information operations with an excitable field. Phys. Rev. E 59, 5354–5360 (1999)
12. Rovinsky, A.B.: Spiral waves in a model of the ferroin catalyzed Belousov-Zhabotinsky reaction. The Journal of Physical Chemistry 90(2), 217–219 (1986)
13. Rovinsky, A.B., Zhabotinsky, A.M.: Mechanism and mathematical model of the oscillating bromate-ferroin-bromomalonic acid reaction. The Journal of Physical Chemistry 88(25), 6081–6084 (1984)
14. Tóth, Á., Showalter, K.: Logic gates in excitable media. J. Chem Phys. 103(6), 2058–2066 (1995)
15. Turing, A.M.: The chemical basis of morphogenesis. Philosophical Transactions of the Royal Society of London. Series B, Biological Sciences 237(641), 37–72 (1952)
16. Maass, W.: Networks of spiking neurons: The third generation of neural network models. Neural Networks 10(9), 1659–1671 (1997)

Extending an Established Simulation: Exploration of the Possible Effects Using a Case Study in Experimental Autoimmune Encephalomyelitis

Richard B. Greaves[1], Mark Read[2], Jon Timmis[1,2],
Paul S. Andrews[1], and Vipin Kumar[3]

[1] Department of Computer Science, Deramore Lane, University of York, UK
{rgreaves,jtimmis,psa}@cs.york.ac.uk
[2] Department of Electronics, University of York, Heslington, York, UK
mnr101@ohm.york.ac.uk
[3] Laboratory of Autoimmunity,
Torrey Pines Institute for Molecular Studies, La Jolla, USA

Abstract. Investigation of a biological domain through simulation can naturally lead to the desire to extend the simulation as new areas of the domain are explored. Such extension may entail the incorporation of additional cell types, molecules or entire molecular pathways. The addition of these extensions can have a profound influence on simulation behaviour, and where the biological domain is not well characterised, a structured development methodology must be employed to ensure that the extended simulation is well aligned with its predecessor. The paper presents such a methodology, relying on iterated development and sensitivity analysis, by extending an existing simulation of Experimental Autoimmune Encephalomyelitis (EAE), a disease model for Multiple Sclerosis, via inclusion of an additional regulatory pathway. We reflect on the implications of extensions which alter simulation behaviour on pre-extension results.

Keywords: Regulatory pathway, Experimental Autoimmune Encephalomyelitis, agent-based simulation, *in silico* experimentation.

1   The Proposed Methodology

Computational simulation is a useful complement to wet-lab techniques, providing a means to conduct, *in silico*, experiments that might be impossible to conduct using current laboratory technology [15]. However, simulation can only ever be based upon some simplification or abstraction of the system of interest, as it is this abstraction that makes the simulation computationally tractable. Moreover, the quality of the system model employed in simulation will be dependent upon the state of knowledge of the real-world domain. Therefore it is a natural expectation that as domain knowledge increases, the working model will need to be enhanced, be that via the addition of new system components or via new modes of interaction between them. Here we

M.A. Lones et al. (Eds.): IPCAT 2012, LNCS 7223, pp. 150–161, 2012.

discuss the possible implications of extending existing simulation and present a methodology for the implementation of such extensions.

In implementing extensions to a simulation, an iterative development procedure which mirrors the CoSMoS methodology [1] is employed. Development commences with the design of a very simple model as there may be little in the literature to guide the development of a more sophisticated model. This developmental phase is guided by the domain expert and seeks to identify the key actors and the relevant modifications to their current behavior in the model. Additions to the basic simulation will necessitate the reparameterisation of the simulation as the previous calibration can no longer be assumed to be valid. Therefore, once the extensions to the simulation are implemented, it is then important to explore the effects of any new simulation parameters introduced to the model. One way of doing this is to systematically map simulation behavior over the full range of values of any parameters introduced to the model, referred to here as **factorial analysis** [17]. However, this will only be tractable if very few new parameters are introduced. If such an initial exploration of parameterization is not feasible, then a fuller exploration of simulation parameterization will be called for, for example using a global **sensitivity analysis** employing an efficient sampling technique, for example Latin Hyper Cube sampling [11]. This analysis will be necessary in any case following assessment of observed effects with the help of the domain expert and will serve to re-balance the effects of the extension parameters with those already included in the model.

Sensitivity analysis is a form of statistical analysis that allows the attribution of variation in system outputs to variation in inputs. Such analysis employs an efficient sampling of parameter space as a systematic mapping of parameter space would be computationally intractable. It is good practice to understand how all parameters introduced by extensions to the model change it. This information is important for reparameterisation of the simulation, which if performed properly, will entail changes to all simulation parameters, including those that are well tied to values from the literature. Abstraction entails that simulation parameters do not represent exactly the same thing as the corresponding *in vivo* values as they must compensate for other pathways and components that are omitted from the model.

Ultimately the simulation ought to be fully recalibrated to establish baseline behaviour. This need to iteratively recalibrate upon extension of a model suggests two important implications for the use of simulation in biology: the need for caution in using the quantitative results of simulation and the need for strong methodology to guide simulation development.

2   The Proposed Case Study

The specific domain of interest for this work is Experimental Autoimmune Encephalomyelitis (EAE), a mouse model of Multiple Sclerosis. This is presented as a case study of on-going research using an Agent-based EAE simulator, ARTIMMUS [13],

[14] which was developed following principled methodology and has been calibrated against *in vivo* data and which we now wish to extend to incorporate more domain detail. ARTIMMUS was developed using the CoSMoS process [1] which provides a principled framework for the development of complex system models and simulation, promoting trust in the simulation and the results emerging from it. In accordance with the CoSMoS process explicit domain modeling was carried out, wherein specific models were created to answer specific questions in a specific domain and evaluations were guided at all times by the domain expert. The simulation has undergone an iterative calibration whereby simulation development is guided by continual improvements in the alignment of simulation predictions and *in vivo* behavior. Hence, the abstractions and assumptions made in simulation are informed by the domain expert and by empirical evidence of their appropriate capture of the domain [13], [14].

Subsequent to its development, ARTIMMUS has also been rigorously tested for simulation sensitivity to perturbations in parameter values [15]. Global **sensitivity analysis** permits the modeller to systematically examine the relative influence of each simulation parameter on simulation behavior. A robustness analysis has been employed which reveals the extent to which simulation predictions are genuinely representative of the domain rather than of underspecified parameter values arising from inconclusive domain knowledge.

An additional regulatory pathway that is believed to be influential in regulating autoimmune disease states has been identified [3]. It is known from the domain expert which cells are involved in this regulatory mechanism, but the pathway is not yet well characterised in the domain. It is therefore the intention to explore the nature of this pathway through simulation, starting by implementing a simplified model and executing exploratory simulations. The synergy of *in silico* and *in vivo* exploration and the resultant discussions with the domain expert should allow for a better characterization of the pathway via generation of experimentally testable hypotheses concerning the nature of the pathway.

The model development strategy utilized here is iterative in nature, the work presented representing an initial exploration of the pathway *in silico*. A very simplified, initial simulation has been created in which further complexity can be incrementally implemented as required since there exists very little practical guidance in the literature as to how the final simulation should behave. Since the baseline behaviour of the simulation has been characterized [14], it will be necessary to modify the parameterisation and / or the implementation of the pathway to reproduce the previously validated baseline behaviour.

The remainder of the paper addresses the disease model employed, how the additions to the simulation were implemented and the results attained. To this end, preliminary detail of the EAE model employed and the extensions to be made to it are provided in Section 3. Section 4 discusses the model of the added regulatory pathway that was implemented and Section 5 presents the results obtained from simulation based on the augmented model. Finally, Section 6 discusses the results and draws conclusions.

3   Experimental Autoimmune Encephalomyelitis

Experimental Autoimmune Encephalomyelitis (EAE) is an autoimmune disease that serves as a model for Multiple Sclerosis [10]. Multiple Sclerosis is characterised by damage to the myelin coating of nerve fibres resulting in impaired conduction of impulses along them, leading ultimately to paralysis and death. In EAE similar damage to the myelin sheath is mediated by CD4 Th1 cells that are reactive towards various components of myelin, for example Myelin Basic Protein (MBP) [19].

This case study is focussed on the murine model of EAE [7, 18]. This model addresses the mechanisms of spontaneous recovery from EAE which is highlighted as grey dashed arrows in Figure 1.

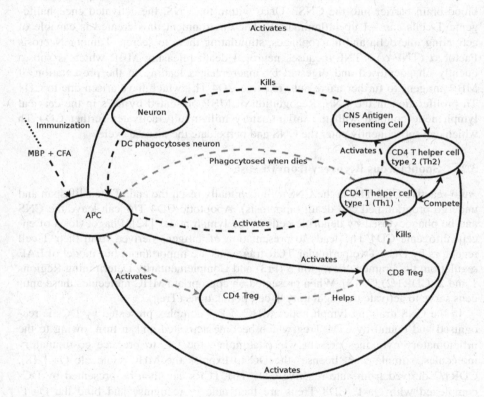

Fig. 1. An informal representation of the cells implicated in the EAE disease and spontaneous recovery cycles and their inter-relationships. The disease cycle is highlighted in black dashed lines, the recovery cycle in grey dashed lines. Dendritic cells (DCs) and macrophages are represented in the cycle as generic Antigen Presenting Cells (APCs). Figure adapted from [16].

3.1   Induction of Disease

The disease cycle is informally represented by the circuit of black dashed arrows in Figure 1. The disease can be induced in laboratory animals via inoculation with MBP, Complete Freund's Adjuvant (CFA) and Pertussis toxin [12], which provokes an immune response and stimulates DCs sufficiently that they can express costimulatory molecules. Dendritic cells (DCs) respond to MBP by phagocytosing (internalizing) and presenting it as complexes with MHC molecules. The DC then migrate to the secondary lymphoid organs where they encounter naïve MBP-recognising CD4 T-helper cells and bind and activate them, allowing the auto-reactive CD4 Th1 to proliferate and mature into effectors.

The activated T-cells can then migrate, along with macrophages, through the blood-brain barrier into the CNS. Once within the CNS, the activated encephalitogenic T-cells can set up inflammation. The environment thus created is capable of activating microglia and macrophages, stimulating them to secrete Tumour Necrosis Factor α (TNF-α). TNF-α causes neuronal death releasing MBP which is subsequently phagocytosed and digested by macrophages leading to the presentation of MBP antigens to further naïve auto-reactive CD4 Th1 which have arisen due to CD4 Th proliferation in the CNS. Recognition of MBP presented by DCs in the cervical lymph node coupled with the inflammatory milieu fully activates further CD4 Th which can subsequently enter the CNS and perpetuate the disease cycle.

3.2   Spontaneous Recovery from Disease

Auto-reactive CD4 Th1 in the CNS will eventually reach the end of their lifespan and undergo programmed cell death (apoptosis). Apoptotic CD4 Th1 can leave the CNS and be phagocytosed by dendritic cells in the lymph nodes [12]. Phagocytosis of encephalitogenic CD4 Th1 leads to presentation of antigens derived from their T-cell receptors by DCs. Two particular TCR fragments are important in our model of EAE regulation: the Framework region 3 (Fr3) and Complementarity Determining Regions 1 and 2 (CDR1/2) [7, 18]. When presented on appropriate MHC molecules these antigens serve to activate the regulatory T-cell populations (Treg).

In the CNS draining lymph nodes MHC-II-Fr3 complex presented by DCs is recognised and bound by CD4 Treg which become activated and in turn, owing to the inflammatory cytokines present which stimulate the DC to produce costimulatory molecules, stimulate or 'license' the DC to express the MHC molecule Qa-1 [8]. CDR1/2 derived from auto-reactive CD4 Th1 TCRs can then be presented by DCs complexed with Qa-1. CD8 Tregs are then able to recognise and bind the Qa-1-CDR1/2 complex and become activated by the DC with help from IFN-γ secreted by CD4 Treg [18]. Once activated, the CD8 Treg leave the DC and are capable of recognising and binding Qa-1-CDR1/2 presented transiently on the surface of auto-reactive CD4 Th1 [7]. CD8 Treg kill these by inducing apoptosis [2]. This cell-mediated killing serves to regulate the auto-immune response and the population of self-reactive CD4 Th1 is reduced, the spontaneous recovery from EAE and the return of cellular populations to their resting levels being essentially complete within 50 days [12]. This behaviour is reflected in the ARTIMMUS results model presented in Figure 2.

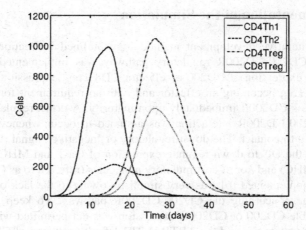

Fig. 2. The populations of T-cell effectors with time. The curves are derived from the median population levels at a given time across a set of 1000 simulator runs. The resulting curves clearly show the anticipated behaviour of the EAE system: cell populations begin at their resting levels at day 0 when immunisation occurs. The populations of the CD4 Th cells then begin to rise, peaking at around 10-15 days, with the CD4 Th1 population peaking considerably higher than CD4 Th2 (circa 1,000 cells against circa 200). By this time, the populations of Tregs will have begun to increase and these will peak around 27-30 days with CD4 Treg attaining a higher population than CD8 Treg (circa 1,000 cells against circa 600). Recovery from EAE will now be effectively complete and cell populations will fall back to their resting levels. The plot was generated by re-running the baseline experiment as described in [16].

3.3   The CD200-CD200R Immune-Regulatory Pathway

The pathway added to the EAE model is thought to reduce auto-immune response by down-regulating the ability of DCs to stimulate T-cell populations. DCs are pivotal in the activation of all T-cell sub-populations and so any reduction of their capacity to bring this about could significantly scale back the size of an immune response by curtailing T-helper and Treg populations.

CD200 and CD200R are cell surface proteins, CD200 being a ligand to the CD200R receptor [5]. CD200 is expressed on a variety of cells particularly T-cells from the immune system and also on neurons [9]. The CD200 protein has also been shown to be expressed constitutively on the CD8 Treg population [4]. CD200R is similarly widely expressed with microglia and DC being able to express this receptor [9]. In the initial augmented model, only T-cell-DC interactions outside the CNS are considered in line with the iterative development methodology. When CD200 binds to CD200R, a signal is transmitted via the receptor to the DC or microglia, causing it to down-regulate production of certain proteins that are essential to the binding and activation of T-cells [5]. There is now considerable experimental evidence for the regulatory effect of CD200 on cells such as DC and macrophages [9].

4   Implementation of the Simulation

Following the iterative development methodology outlined in Section 1, a simple model of the CD200-CD200R regulatory pathway was implemented. The model allowed for the expression of CD200 by effector CD8 Treg, expression being immediate upon the Treg becoming an effector and with no requirement for local activation. DCs express CD200R immediately upon maturity. A down-regulatory signal to DCs from CD200-CD200R interaction was assumed to occur whenever CD8 Treg and DCs come into contact. The down-regulatory or 'negative' signal then probabilistically causes the DC to down-regulate expression of Qa-1 and MHC (collectively referred to as MHC) and / or of costimulatory molecules (referred to as CoStim).

The model implemented [6] was very simplistic owing to the lack of detailed explicit knowledge concerning the CD200-CD200R pathway. To keep the model as simple as possible CD200 or CD200R expression was not permitted within the CNS i.e on neurons or microglia and the ARTIMMUS default exclusion of CD8 Treg from the CNS [14] was maintained.

The stipulation that negative signalling probabilistically down-regulates MHC and / or CoStim expression by DCs introduces two new parameters into our model, namely the probability that a negative signal will down-regulate MHC expression and also the probability that negative signalling will down-regulate CoStim expression by DCs. We therefore seek to ascertain valid values for these new parameters, such that the added pathway will not unduly perturb simulation from its baseline behaviour. This is done with a view to re-balancing the effects of the CD200-CD200R regulatory pathway with those of the regulatory pathway already implemented in the simulation.

Since the pathway is poorly characterised it is not possible to obtain reasonable values for the probabilities from the literature. The effect of the existing simulation parameters will also be slightly perturbed by the additions to the model. These considerations therefore make it necessary to conduct a full analysis of the effects of the parameters introduced by extending the model. A factorial analysis i.e a systematic mapping was conducted of simulation behaviour across the full range of values for both of the added parameters.

5   Evaluation of the Initial Model of the CD200 Pathway

The **factorial analysis** mapped simulation behaviour for values of the probability parameters between 0% and 100% in steps of 10%, meaning a mapping of 121 separate simulations with unique pairings of probability values. Simulation behavior was described in terms of median values across 1000 simulator runs for various system properties or responses. The responses analysed were peak T-cell effector populations (for CD4 Th1, CD4 Th2, CD4 Treg and CD8 Treg), the times taken to reach these maximal populations, the CD4 Th1 population remaining at day 40 of the simulation, the mean EAE severity score and the EAE severity score at 40 days. EAE severity scores calculated from effector cell populations are intended to correspond to the clinical severity scores assigned in the laboratory on the basis of the observed extent of disease symptoms [14].

Typically, even when there was only a 10% chance that MHC expression could be reduced via negative signalling; the effect on T-cell effector populations was significant. CD8 Treg and CD4 Treg peak populations were significantly reduced compared to the experiment with both parameters set to 0% i.e. the baseline (data presented in Figures 3 and 4 respectively). The peak population of CD4 Th1 was increased, though not significantly, presumably owing to the reduced activation by DC being balanced by the reduction in the occurrence of apoptosis by CD8 Treg.

Repetition of the factorial analysis with values for the probability parameters between 0% and 1% in steps of 0.1% still showed significant reductions in CD4 Treg and CD8 Treg populations compared to baseline values even at values of the MHC down-regulation probability as low as 0.2%.

The **factorial analysis** highlights a need to further develop our model of the CD200-CD200R regulatory pathway, the existing model exerting too drastic an effect on T-cell population priming and almost certainly not being a fair representation of the domain. In practice, the model leads to a severe reduction in T-cell effector populations, which is believed to be excessive. This arises from the impact that CD200 expressed by Treg has on all T-cell populations (illustrated informally in Figure 5). It is difficult to establish exactly how one should go about implementing such a sparsely characterised pathway in the model, because of the potentially far reaching consequences of reducing the capacity of DCs to prime T-cells in this particular instance. Via the expression of CD200, CD8 Treg can impact not only the priming of all other T-cell sub-populations, but also their own priming by DCs.

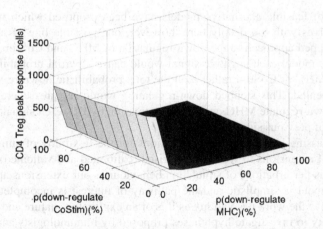

Fig. 3. Three-dimensional plot of peak CD8 Treg effector populations in a systematic mapping of simulation behaviour at values of the two probability parameters between 0% and 100% in steps of 10%. The probability parameters represent the probability that a negative signal will down-regulate MHC (i.e. MHC-II and Qa-1) expression by a DC – represented as p(down-regulate MHC) on the axes and the probability that a negative signal will down-regulate CoStim expression by DC – represented as p(down-regulate CoStim) on the axes. 121 parameter value pairings were used in generating the landscape, the baseline simulation is represented as the pairing 0%, 0% in the bottom middle of the plot.

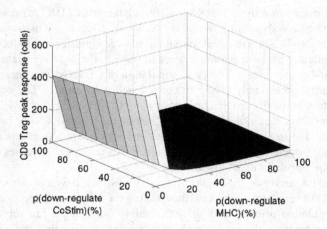

Fig. 4. Three-dimensional plot of peak CD4 Treg effector populations in a systematic mapping of simulation behaviour at values of the two probability parameters between 0% and 100% in steps of 10%. The probability parameters represent the probability that a negative signal will down-regulate MHC (i.e. MHC-II and Qa-1) expression by a DC – represented as p(down-regulate MHC) on the axes and the probability that a negative signal will down-regulate CoStim expression by DC – represented as p(down-regulate CoStim) on the axes. 121 parameter value pairings were used in generating the landscape, the baseline simulation is represented as the pairing 0%, 0% in the bottom middle of the plot.

A number of feasible alternative models have been proposed which we shall not concern ourselves with too deeply here, however, one possible line of investigation concerns a stepped approach to the down-regulation of MHC and CoStim by DCs. In this alternative model each negative signal would cause a partial probabilistic down-regulation of MHC or CoStim rather than the total probabilistic down-regulation currently implemented. This stepped down-regulation would require several signals to completely down-regulate MHC or CoStim expression by DC, each signal compounding the effect of previous signals.

Of more pressing import are the implications for the flexibility of simulation as a research tool if extensions to well characterised (calibrated and validated) simulation result in serious perturbations of simulation behaviour as our extensions appear to do. Our current model is simplistic and the pathway of interest is incompletely understood. We believe the work has validity as it is of an exploratory nature and its strength lies in its ability to investigate hypotheses proposed by immunologists and feed back to them the results of simulating this system. In this way, we aim to help shape immunological thinking and thus guide wet-lab experimentation.

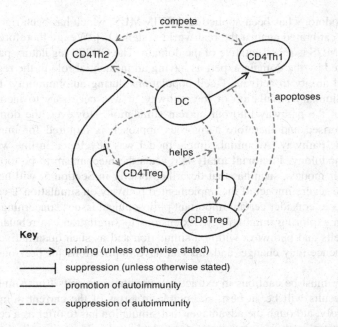

Fig. 5. An informal description of the interactions of the different T-cell populations of our model with each other and with dendritic cells (DCs). Via CD200 expression, CD8 Treg can modulate DC ability to prime all T-cell populations, including CD8 Treg and can this significantly impact T-cell effector populations. Interactions marked as solid black lines tend to promote autoimmunity, whereas those indicated as grey broken lines serve to reduce it. Figure reproduced from [14].

6   Discussion and Conclusions

The paper presents a methodology which is applicable to the extension of established models in order to explore new pathways and components. In this instance, EAE has been employed as a case study, entailing the extension of the ARTIMMUS simulation via addition of a model of the CD200 pathway which considers only the DC-CD8 Treg interactions. The resulting simulation behavior leads us to consider the wider implications of extending simulations in computational immunology.

Simulation represents a means of integrating biological data. A parameterized simulation, once properly calibrated, can serve as a tool for formulating and testing hypotheses relating to the domain. In the natural course of research it is possible that new pathways or components will be identified as being influential in the system, and simulation provides a means whereby a preliminary exploration of these may be conducted. However, as often little is known of the domain, a structured and principled approach to this exploration is required. An appropriate methodology for such exploration has been presented here.

The methodology has been applied to ARTIMMUS, which has been rigorously developed and calibrated against the real-world system [15]. We are therefore confident that ARTIMMUS is representative of the domain. The CD200 regulatory pathway has been identified by the domain expert as having an influential role in the regulation of dendritic cell ability to activate T-cell populations during autoimmunity. In order to conduct preliminary simulation of the pathway, it was necessary to incorporate an abstraction of the pathway into our current simulation. However, the domain is not well characterised and therefore a rigorous approach is required for implementing models of the pathway. An initial, simple model was developed in line with the described methodology. **Factorial analysis** reveals the mechanism to be too simplistic and so further models, such has that described briefly in Section 5, will be explored. However, the severe impact of the implemented pathway on simulation T-cell populations leads us to consider certain important philosophical issues concerning the use of simulation in exploring immunology. Extensions to simulation can rebalance the influence of cells and pathways within a simulation and as such quantitative measures of their influence may change, and this can have implications for previous quantitative results.

Firstly, we must be cautious in extracting quantitative results from simulation and qualitative results will be, at best, acceptable in terms of the current domain knowledge. Secondly, although the advantages that simulation has to offer as a complement to wet-laboratory experimentation are formidable, it is, however, clear that strong methodologies are required to guide simulation development to appropriate levels of abstraction. These methodologies will also serve to ensure that simulation adequately captures the domain, that modellers continually evaluate simulation performance against real domain observations and that simulation is capable of indicating when something influential, and possibly as yet unidentified *in vivo*, is missing from the simulation. A failure of simulation to fully represent real-world dynamics should lead to further *in vivo* exploration or development of the simulation.

Simulation has the potential to offer a powerful complement to *in vivo* study. However, for that potential to be realized, the simulation community needs to employ methodologies that instill confidence that simulation results are accurate and can clearly motivate further development (of the simulation) when this is not the case.

Acknowledgement. We wish to acknowledge Dr Carl Ritson for allowing us access to the CoSMoS computing cluster at the University of Kent in order to carry out the simulation reported here. Mark Read is funded by FP7: ICT grant "Collective Cognitive Robotics (CoCoRo)". Paul Andrews is funded by EPSRC grant "Resilient Futures" grant number EP/I005943/1.

References

[1] Andrews, P.S., Polack, F.A.C., et al.: CoSMoS Process: A Process for the Modelling and Simulation of Complex Systems, University of York, Technical Report YCS-2010-453 (2010)

[2] Beeston, T., Smith, T.R.F., et al.: Involvement of IFN-g and perforin, but not Fas/FasL interactions in regulatory T cell-mediated suppression of experimental autoimmune encephalomyelitis. Journal of Neuroimmunology 229, 91–97 (2010)

[3] Feuer, R.: Tickling the CD200 Receptor. A Remedy for those Irritating Macrophages. The American Journal of Pathology 171, 396–398 (2007)

[4] Fanchiang, S.S., Cojocaru, R., et al.: Global expression profiling of peripheral Qa-1 restricted CD8aa+TCRab+ regulatory T cells reveals innate-like features: implications for immune-regulatory repertoire. Human Immunology (2011) (in Press)

[5] Gorczynski, R.M., Chen, Z., Kai, Y., Lee, L., Wong, S., Marsden, P.A.: CD200 is a Ligand for All Members of the CD200R Family of Immunoregulatory Molecules. Journal of Immunology 172, 7744–7749 (2004)

[6] Greaves, R.: Computational Modelling of Treg Networks in Experimental Autoimmune Encephalomyelitis, MSc by Research Thesis, Department of Computer Science, University of York (2011)

[7] Kumar, V., Sercarz, E.: An Integrative Model of Regulation Centred on Recognition of TCR / MHC Complexes. Immunological Reviews 182, 113–121 (2001)

[8] Kumar, V., Stellrecht, K., et al.: Inactivation of T Cell Receptor Peptide-specific CD4 Regulatory T Cells Induces Chronic Experimental Autoimmune Encephalomyelitis (EAE). Journal of Experimental Medicine 184, 1609–1617 (1996)

[9] Liu, Y., Bando, Y., et al.: CD200R1 Agonist Attenuates Mechanisms of Chronic Disease in a Murine Model of Multiple Sclerosis. Journal of Neuroscience 30, 2025–2038 (2010)

[10] Lublin, F.D.: Relapsing Experimental Allergic Encephalomyelitis an Autoimmune Model of Multiple Sclerosis. Springer Seminars in Immunopathology 8, 197–208 (1985)

[11] McKay, M.D., Beckmann, R.J., et al.: A Comparison of Three Methods for Selecting Values of Input Variables in the Analysis of Output from a Computer Code. Technometrics 21, 239–245 (1979)

[12] Pender, M.P.: Experimental Autoimmune Encephalomyelitis. In: Pender, M.P., McCombe, P.A. (eds.) Autoimmune Neurological Disease, pp. 26–88. Cambridge University Press, Cambridge (1995)

[13] Read, M., Andrews, P., et al.: Nature Biotechnology (2011) (in preparation)

[14] Read, M.: Statistical and Modelling Techniques to Investigate Immunology Through Agent-based Simulation, PhD Thesis, Department of Computer Science, University of York (2011)

[15] Read, M., Andrews, P., et al.: Techniques for Grounding Agent-based Simulations in the Real Domain: a Case-study in Experimental Autoimmune Encephalomyelitis. Mathematical and Computer Modelling of Dynamical Systems (2011), doi:10.1080/13873954.2011.601419

[16] Read, M., Timmis, J., et al.: A domain model of experimental autoimmune encephalomyeltis. In: Stepney, S., Welch, P.H., Andrews, P.S., Timmis, J. (eds.) CoSMoS, pp. 9–44. Luniver Press (2009)

[17] Saltelli, A., Chan, K., et al.: Sensitivity Analysis. Wiley series in probability and statistics. Wiley (2000)

[18] Tang, X., Smith, T.R.F., et al.: Specific Control of Immunity by Regulatory CD8 T Cells. Cellular and Molecular Immunology 2, 11–19 (2005)

[19] Van den Bark, A.A., Gill, T., et al.: A Myelin Basic Protein-specific T Lymphocyte Line that Mediates Experimental Autoimmune Encephalomyelitis. Journal of Immunology 135, 223–228 (1985)

Human Uterine Excitation Patterns Leading to Labour: Synchronization or Propagation?

Eleftheria Pervolaraki and Arun V. Holden

Institute of Membrane and Systems Biology
University of Leeds, Leeds LS9 9JT
a.v.holden@leeds.ac.uk

Abstract. The mechanisms leading to the initiation of normal, premature or dysfunctional human labour are poorly understood, as animal models are inappropriate, and experimental studies are limited. Computational modelling provides a means of linking non-invasive clinical data with the results of *in vitro* cell and tissue physiology. Nonlinear wave processes – propagation in an excitable medium – provides a quantitatively testable description of mechanisms of premature and full term labour, and a view of changes in uterine electrophysiology during gestation as a trajectory in excitation and intercellular coupling parameter space. Propagation phenomena can account for both premature and full term labour.

Keywords: Computational biology, uterus, pregnancy, birth.

1 Introduction

The initiation of normal human labour is poorly understood and, because of rapid evolutionary divergence, and the problems associated with the combination of the pelvic geometry adapted for a bipedal gait, and the large skull size, [Franciscus 2009] insights from experimental animals are of limited relevance. Unlike birth in intensively studied common laboratory animals (*e.g.* mouse, rat, rabbit) or farm animals (e.g. sheep, pigs), human birth is not preceded by dramatic fall in plasma progesterone, and the trigger for it is not known. Multiple interactions between human labour, foetal and placental membrane, and maternal signalling lead to maternal and foetal oxytocin, as well as placental oestriol and prostaglandlins acting on receptors on uterine smooth muscle (myometrial) cells [Norwetz *et al.* 1999]. Normal parturition requires appropriate positioning of the foetal head (engagement) within the pelvis, cervical ripening in addition to coordinated uterine contractions. However, uterine contractions can be considered a final common pathway for the mechanisms controlling parturition. These contractions not only lead to the delivery of the baby and placenta, but after delivery the uterus contracts down from an organ length of ~30cms to ~7cms, minimising blood loss through haemorrhage. Although most births proceed normally, disorders in the timing and pattern of uterine contraction occur.

Premature contractions can lead to preterm birth *i.e.* birth before 37 weeks of gestation. About 10% of births are preterm, and preterm birth is a direct cause of over

M.A. Lones et al. (Eds.): IPCAT 2012, LNCS 7223, pp. 162–176, 2012.

one million neonatal deaths/year worldwide. In developed countries, preterm birth is the single greatest contributor to neonatal mortality and morbidity, accounting for 50% of infant neurological damage, and can lead to life-long morbidity. In addition to the devastating health, social and emotional effects that preterm birth has on babies, parents and the extended family, there are extensive financial costs to families and society. Spontaneous preterm labour (*i.e.* inappropriate early activation of uterine contractions) is the single biggest cause of preterm birth with data showing rising rates [Norman *et al.* 2009].

Ineffective myometrial contractions before delivery is the second leading cause of Caesarean section rates, which in developed countries are around double the rate recommended by the World Health Organisation, and are associated with significant increased maternal mortality and morbidity in addition to extra health care costs. Obstetric haemorrhage is largely due to failure of the myometrium to contract effectively after delivery, and is the leading cause of maternal death worldwide accounting for over 132,000 maternal deaths annually.

Understanding the mechanisms of initiation and control of human myometrial activity during pregnancy and labour requires human data. During a Caesarean delivery, biopsies of uterine myometrium are available, so *in vivo* experiments on (near) full term human myometrial tissue and cells are feasible. During pregnancy, the foetus and uterus are routinely imaged by ultrasound, monitoring uterine growth and foetal development. Occasionally there is a clinical need for magnetic resonance imaging (MRI), and so uterine geometry can be reconstructed in three dimensions. The electrical activity of the uterus can be extracted by filtering out foetal and maternal ECG components from non-invasive electrophysiological recordings with electrodes on the surface of pregnant abdomen.

We aim to improve our understanding on the control of uterine organ level activity during human pregnancy and labour by virtual uterine tissue engineering – the construction and solution of biophysically and anatomically detailed computational models of uterine cell, tissue and organ electromechanics, combined with time series of uterine activity recordings during pregnancies [Aslanidi *et al.*, 2011]. This considers the behaviour of the uterus as the behaviour of an excitable medium, where synchronised activity – the contractions of labour – could be driven by a specialised pacemaker site (as in the heart), or emerge *via* synchronisation of weakly coupled oscillators, or by the propagation of endogenous nonlinear waves. These possible alternative control strategies depend on the excitation and coupling parameters of the medium, that change during gestation, and understanding the control strategy would help to target interventions aimed at dysfunctional contractions. If these interventions are pharmacological then detailed, quantitative models of USMC membrane excitation are necessary.

2   Uterine Smooth Muscle Cell Electrophysiology and Models

There are a number of detailed biophysical models for the electrophysiology of uterine smooth muscle cells (USMCs) – these are "mammalian" i.e. chimaeric, based on voltage clamp data from different species, predominantly the laboratory rat [Bursztyn et al., 2007; Rihana et al., 2009; Tong et al., 2011]. These all relate the electrical activity of an isolated cell V(t) to the sum of membrane ionic current I_{ion} flows through pumps, exchangers and channels; some of which are voltage and/or time dependent, and have a structure as in Fig. 1:

$$C_m \, dV/dt = \Sigma \, I_{ion}. \tag{1}$$

A recent model of a USMC has 13 ionic currents and simple calcium dynamics and is described by 25 dynamical variables and >125 parameters. The currents include:

Inward currents: L-type Ca^{++} current (I_{CaL}), attributed to be the major inward current, fast Na+ current (I_{Na}), T-type Ca^{++} current (I_{CaT}) and a hyperpolarisation-activated current (I_h). A nonspecific cation current (I_{NSCC}) is also included, with a reversal potential within the reported AP amplitude range.

Outward currents: Fast A-type transient K^+ current (I_{Ka}), two voltage-gated K^+ currents (I_{K1} and I_{K2}), Ca^{++} - activated K+ current ($I_{K(Ca)}$) and a leak background current (I_{Kleak}).

Electrogenic exchangers: Na+- Ca^{++} exchanger current (I_{NaCa}) and Na^+-K^+ pump current (I_{NaK}).

Calcium fluxes: The $[Ca^{++}]_i$ dynamics, the Na^+- Ca^{++} exchanger (J_{NaCa}) and the plasma membrane Ca^{++} -ATPase (J_{PMCA})) is modified from a simple uterine excitation-contraction model to include the kinetics from membrane calcium channels and electrogenic I_{NaCa}.

Numerical solutions of the USMC model show spiking and plateau action potentials, and can be modified to reproduce the effects of hormones *via* its action on the magnitude and kinetics of Ca^{++} and K^+ currents, and reproduce V(t), $[Ca^{++}]_I$ and force under a variety of conditions (see Fig. 2). The excitability and action potentials of mammalian USMCs from the pregnant uterus change during gestation and with changes to the foetal location within the uterus. Parameters in the cell model can be modified to reproduce these changes in uterine electrophysiology. In vitro electrophysiological recordings can be performed on human myometrial cells and tissues from the lower segment of the uterus obtained at Caesarean sections (i.e. close to full term). These changes in electrophysiology are due to up/down regulatory changes in channel expression – e.g. the slow drift in human USMC membrane potential from ~-75mV at 28 weeks to ~-55 mV at full term [Partington et al. 1999] can be accounted for by a slow upregulation of K^+ selective conductances.

During gestation the state of the USMC, defined by its set of parameter values, may be considered to be moving in parameter space. An extensive survey of the electrophysiology of human USMCs during gestation is not possible, but the presence and approximate magnitudes of different conductance systems can be estimated from immunohistochemistry and mRNA mapping of tissue samples obtained. The kinetics of the different ionic channels types can be obtained from single channel patch clamp

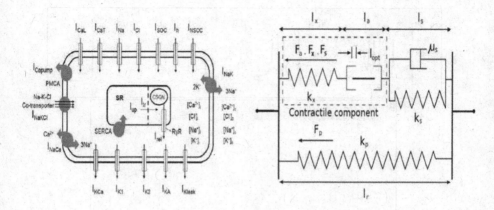

Fig. 1. A schematic representation of (a) the ionic channels, pumps and exchangers in a model for a uterine smooth muscle cell (USMC). Changes in $[Ca^{2+}]_i$ follow membrane Ca^{2+} fluxes, with sarcoplasmic reticulum SR with separate uptake (I_{up}) and release (I_{rel}) compartments. Ca^{2+} is transferred to the SR through the SERCA pump, Ca^{2+} is then diffuses across (I_{tr}) to the release compartment before releasing back to the cytosol through the ryanodine receptor flux. The $[Ca^{2+}]_i$ drives tension development in the mechanical component model based on Hill's three component model where the F_{tot}: total force = F_a: active force + F_p: passive force, and the active force is generated by cross bridge interactions . $F_x=F_a=F_s$. F_x: cross bridge elastic force; F_s: series visco-elastic force lc: total cell length; la: length of the active contractile component lx: extension length of the cross bridge; ls: length of passive component.

studies on human channels expressed in *Xenopus* oocytes or mammalian expression lines, and so human USMC models, based on data from human cells and tissues, are being constructed. Isolated mammalian USMCs show a wide range of electrophysiological activity, from spike like action potentials, repetitive spiking, action potentials with long plateaux and bursts of spikes on the plateau. USMC models need to reproduce these different behaviours, in different regions of parameter space. Bifurcation analysis of USMC models, *via* the use of continuation algorithms, identifies equilibrium solutions, their stability, and perhaps bistability; periodic solutions and their stability, and how these interact to produce bursting activity (Fig. 3).

Trafficking, channel turnover and sequestration within caveolae also influence the density of the ion channels on the cell surface, and so the membrane parameters of the USMC model may fluctuate. Bifurcation analysis allows the characterisation of complex bursting patterns in terms of how they arise dynamically, rather than by what individual ionic mechanisms contribute to their generation [Bertram et al, 1995]. A qualitative understanding of the bifurcational structure of the USMC model in parameter space provides a view of how the excitation dynamics can change during gestation, and a detailed numerical bifurcation analysis of specific USMC models can quantitatively account for specific experimental results *e.g.* the effects of oestradiol

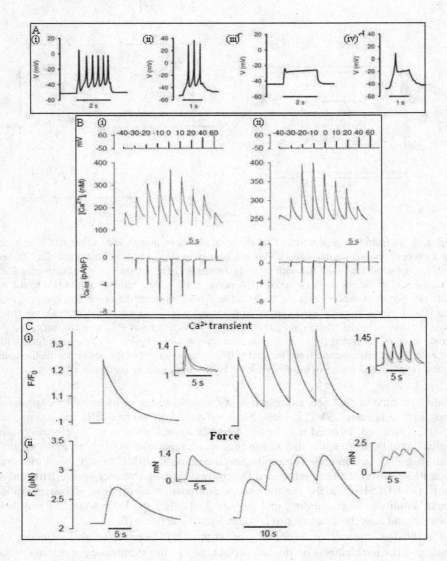

Fig. 2. Numerical solutions of USMC model and experimental recordings. (A) Electrophysiology in physiological saline (i, ii), and in presence of oestradiol (iii, iv). (ii, iv) from Inoue *et al.*, 1999. (B) The response to a series of depolarising clamps, with $[Ca^{++}]_i$ and total I_{Ca}: experiments from Shmigol *et al.*, 1998 on left, numerical solutions on right. (C) $[Ca^{++}]_i$ and force in response to spike like action potentials, experimental   insets from Burdyga *et al.*, 2009.

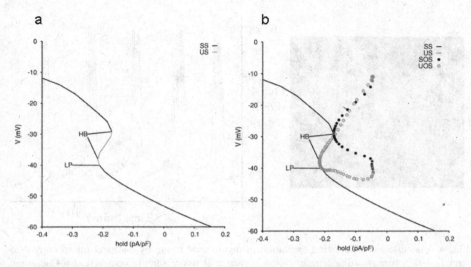

Fig. 3. Bifurcation diagram for the uterine smooth muscle cell model as a constant external current (hold) as the bifurcation parameter is varied. The bifurcation diagram shows different system behaviours, changing from a low resting stable at 50mV, *via* two bistable regions to a more depolarised steady state at 20mV as the bifurcation parameter changes; a) Equilibrium Solution: the stability is lost at a Hopf bifurcation HB and a limit point LP, b) numerical tracking of stable and unstable steady (SS and US) and oscillatory (SOS, UOS) solutions. The bifurcation parameter "hold" represents an applied current, for example a stretch activated current.

3   Emergence of Tissue Electrophysiology

Biopsies of lower uterine segment myometrial tissue obtained at Caesarean section provide single cells for whole cell electrophysiology experiments, as in Shmygol et al. [2007]. In tissue experiments, optical imaging of $[Ca^{++}]_i$ in thin (200 μm) myometrial slices loaded with Ca^{++} sensitive Fluo-4 provides a method of visualizing the spatiotemporal information of excitation, as myometrial $[Ca^{++}]_i$ dynamics are determined by Ca^{++} entry through voltage gated calcium channels and hence reflect changes in membrane potential. When the same methodology is applied to rat myometrial cells and tissues, isolated cells show spiking, action potentials with plateau, and complex action potentials with bursting spikes on the plateau. However, in human isolated smooth muscle cells only spikes and plateau action potentials are observed; complex action potentials (similar to those illustrated in Fig. 4B) are only seen in tissue preparations, and are abolished by gap junction blockers. In the human the complex action potentials that produce synchronous contractions of labour emerge at the tissue level [Bru-Mercier et al., 2007].

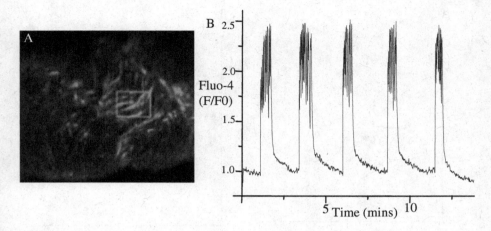

Fig. 4. Visualisation of bursting excitation in myometrial tissue by confocal microscopy of a stretched (1.5 times resting length) human myometial tissue sample obtained at a Caesarean section. (A) A single frame from a movie showing Fluo-4 signal from a thin (~200μm) slice; (B) Time series of calcium bursts as the total sum of fluorescent signal within a selected box. The bursts provide an index of the electrical activity.

4   Connexins, Coupling and Conduction Velocity

The control of muscle activity can be neurogenic, as in skeletal muscles, where motor-neuronal activity initiates, *via* neuro-muscular transmission, muscle excitation, or myogenic, where excitation can be driven by a pacemaker region, or in response to stretch, or myogenic activity modulated by neuronal efferent activity, as in the heart. Coordinated myogenic activity in a tissue requires intercellular coupling: in cardiac and smooth muscle tissue this is *via* gap junctions, composed of connexins, that provide low resistance pathways between cells. During most of gestation the intercellular coupling is weak, and there is an increase in gap junctional expression that precedes normal birth. In laboratory animals this increased connexion expression is regulated by the rising oestrogen and falling progesterone levels. If there was no gap junctional coupling between the USMCs, ephaptic coupling, where a small fraction (membrane resistance >>> extracellular resistance) of the extracellular current produced by activity in one cell could produce very weak coupling between neighbouring USMCs, allowing synchronisation of oscillatory activity but not propagation of waves. Gap junctional coupling provides low resistance pathways that can allow propagation of activity.

An excitable medium modeled by a reaction diffusion system provides a simple continuum model for a tissue composed of coupled excitable cells:

$$\frac{\partial V}{\partial t} = \nabla(D\nabla V) - I_{\text{ion}} \cdot \qquad (2)$$

Here V (mV) is the membrane potential, ∇ is a spatial gradient operator; t is time (ms). D is the diffusion coefficient tensor $(mm^2 \ ms^{-1})$ that characterizes electrotonic spread of voltage. I_{ion} is the total membrane ionic current density $(\mu A \ \mu F^{-1})$ described as in Tong et al. [2011]. D is a measure of the electronic effects of the gap junctional density. For a 1-dimensional excitable model, for a particular I_{ion} , if D is less than some critical value D_c i.e. weak coupling, an action potential cannot propagate, for $D > D_c$ the velocity v_0 of a solitary waves increases as the square root of D. The velocity of a solitary wave solution increases linearly with the magnitude of the fast inward current of (1) and the maximum dV/dt of the upswing of the action potential. For periodic wave trains the conduction velocity decreases at higher rates, as one action potential propagates into the refractory wake of the preceding action potential: this is conduction velocity restitution.

In two dimensions the propagation velocity can be anisotropic, faster in one direction than another, and in a homogeneously anisotropic two-dimensional medium the response to a localised suprathreshold excitation is an ellipsoidal, not a circular, travelling wave, with the long axis of the ellipsoid orientated along the predominant fibre direction. The velocity v of the wavefront also depends on its curvature $k = 1/r$, where r is the local radius of curvature, by the eikonal relation

$$v(k) = v_0 - kD, \qquad\qquad (3)$$

where v_0 is the velocity of a solitary plane wave. In three dimensions there are two principal radii of curvature. The spatio-temporal pattern of propagation of an electrical excitation wave into a resting excitable medium is determined by geometry, rather than details of excitability, and so can be modeled as in Fig. 8 using a simple caricature of excitation. If the wave is propagating into the after-effects of previous activity i.e. into partially recovered, refractory tissue, the spatio-temporal details of the excitation dynamics will influence the pattern of propagation, and so a numerical solution of (2) with a biophysically detailed description of I_{ion} is necessary. The possible spatio-temporal patterns of activity in the uterus depends on its excitability, anisotropic coupling, and geometry, all of which change during gestation.

Conduction velocities can be estimated using invasive, intrauterine electrodes [Wolf and van Leeuwen, 1979], and from non-invasive multi-channel recordings of uterine electrical activity from the abdominal surface. These propagation velocities are estimated from correlation lags or delay times between pairs of electrodes, and so are estimates of propagation velocity only if the wavefront is orthogonal to the lines between the electrodes. At full term the velocity ranges from 2.5 to 5cm/s. For a complex action potential lasting 50s with a conduction velocity of 5 cm/s, the "wavelength" (product of velocity and duration) would be 250cms i.e. larger than the size of the uterus, so a single travelling wave propagates as its wavefront, and would produce a spatially uniform contraction.

If endogeneous uterine excitation was initiated at a consistent pacemaker site (analogous to the sino-atrial node of the heart), or at a specific site (say the entry of the fallopian tube, at the upper "corner" of the uterus) due to boundary conditions then the direction of propagation would be consistent. Propagation from the upper towards the lower segment might be expected. Non-invasive multi-channel electrophysiological recordings do not show any such preferred direction of propagation.

The conduction velocity is determined by the diffusion coefficient in equation (2), which includes the effects of the gap junctional coupling between USMCs. If the intercellular coupling is very low, there is no propagation; as the gap junctional coupling increases propagation becomes possible, and the propagation velocity increases as the square root of the gap junctional coupling conductance. The very low gap junctional expression during pregnancy would prevent propagation; a small increase in gap junctional expression would allow propagation of slow velocity short wavelength waves, and perhaps even re-entrant waves.

5   Time Series Recordings

Electrohysterography (EHG) is the non-invasive monitoring of uterine electrical activity (which are weaker than a few mV and distributed over 0.01–3 Hz) *via* electrodes placed on the abdomen. Since it is non-invasive, the EHG can be used to monitor uterine activity throughout pregnancy and into labour, as a research tool, and to provide a quantitative, objective measure of uterine activity for the management of labour. Changes in the relative power in two frequency bands (0.2-0.45 Hz and 0.8-3Hz) have been described with the progression of pregnancy. Bursts lasting several minutes can be correlated with the timing of contractions. The use of multiple electrodes for EHG recordings allows conduction velocities to be estimated. These recordings of electrical activity during pregnancy also provide maternal and foetal heart ECGs, and the bursts of activity correlate with recorded and perceived contractions.

6   Uterine Tissue Geometry and Architecture

Current clinical imaging of the foetus inside the uterus is *via* ultrasound, to assess the development of pregnancy. This occasionally leads to a clinical need for magnetic resonance imaging that provides high resolution, three-dimensional geometry, from which the geometry of the uterus can be extracted.

Clinical imaging of a full term pregnancy is illustrated in Fig. 6, together with an extracted uterine geometry. Clinical MRI of a full term in vivo uterus on a 3T system does not provide sufficient resolution for reconstructing the architecture of the myometrium. This is necessary for understanding how complex action potentials emerge from the myometrial network, and lead to effective contractions. The detailed architecture has been obtained from *ex vivo* MRI of tissue blocks of a uterus that has been removed after delivery. Diffusion Tensor MRI (DTMRI) provides an index and measures of the myometrial fibre or fibre bundle organisation. The uterus visualised in Fig.6 was removed by hysterectomy after delivery and it is reconstructed in Fig. 6f. Immediately after delivery by a Caesarean section the uterus was flaccid, and had the same size as before delivery; injection of syntocin caused the uterus to contract down to ~9cms, as illustrated in Fig 6(f). After a normal vaginal delivery oxtocin (released in response to the infant   suckling), or syntocin injection by a midwife triggers such a

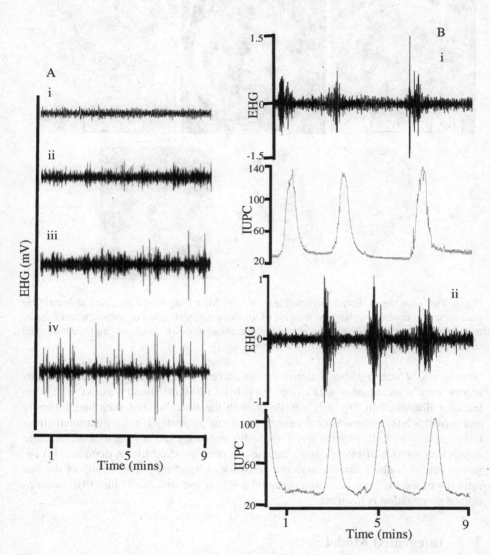

Fig. 5. Non-invasive abdominal surface recordings of uterine electrical activity. A) Recordings during pregnancy at (i) 28, (ii) 32, (iii) 37 and (iv) 40 weeks of gestation. (B) Simultaneous recordings of uterine electrical activity (above) and intra-uterine pressure (IUPC), measured *via* a catheter during (i) early labour, cervix 4 cm dilated; and (ii) late labour, cervix 10 cm dilated.

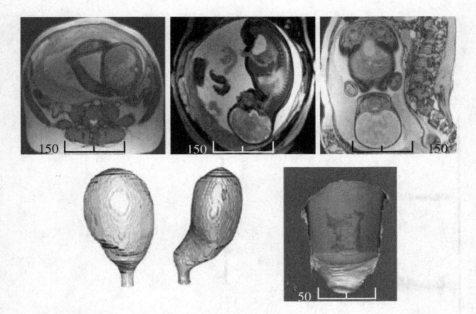

Fig. 6. Full term uterus as reconstructed from *in vivo* MRI images of a pregnant abdomen; (a) transverse, (b) coronal, (c) sagittal images; (d, e) surface visualisation of reconstructed geometry of pregnant uterus (f) *ex vivo*, post partum contracted down uterus of (a-c) reconstructed from MRI. Note massive thickening of myometium.

contraction, which prevents extensive post-partum haemorrhage. Segments of this uterus were scanned on a small bore 9.4T MRI (200μm cuboid voxels). The fibre tracking illustrated in Fig. 7 is consistent with the fibre bundles seen histologically and provides a three-dimensional anisotropic tissue geometry for computational simulation of tissue electrophysiology. Within this geometry the myometrial architecture appears as a series of interweaving bundles, several hundred μm in diameter. Propagation spread is along and through this feltwork of bundles, but the width of the fast upswing of the myometrial action potential is ~1cm and so a continuum PDE description of propagation is justified.

7   Integrated Model

During most of gestation and early labour the uterus is quiescent, and so any propagation is into resting tissue, and so is determined primarily by the geometry, curvature and eikonal relation (equation 3). Simplified models that caricature excitation may be used to explore the possible emergent spatiotemporal patterns of excitation in the uterine geometry; it is only necessary to have appropriate space and time scales i.e. wavelength of excitation, uterine size, and propagation velocity. Fig. 8 illustrates idealised patterns of activity on the uterine surface computed for the isotropic.

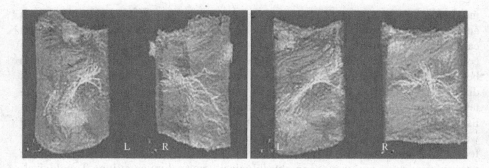

Fig. 7. Diffusion Tensor MRI (DTMRI) visualisation of the architecture of the wall of a human uterus removed after delivery by hysterectomy; A, B) illustration of the tracking of the orientation streamlines of the primary eigenvector field in tissue segments cut in half, representing the left and the right of the same tissue block. These streamlines track the fibre and bundle orientation.

geometry of the in vivo uterus of Fig. 6 using a FitzHugh-Nagumo two-variable caricature for excitation, with standard parameter values – see Winfree [1991] for details of the excitation model. Although the uterus is three-dimensional, it is thin walled, which motivates the assumption that propagation is dominated by the one- and two-dimensional effects of the rate and curvature dependence of propagation velocity. The parameters are for an excitable medium, not an autorhythmic medium, and so activity has been initiated by stimulation.

In Fig. 8(a) excitation is initiated in the fundus, producing an annular wave front propagating in the homogenous, isotropic medium towards the cervix: this represents the trigger for a single contraction. Even though the model is isotropic and homogenous, there are irregularities in propagation velocity due to the uterine geometry and curvature effects. These irregularities are enhanced by static changes in the diffusion coefficient, that simulate spatial granularity in intercellular coupling, or by use of a diffusion tensor obtained from diffusion tensor imaging and that represents anisotropy in intercellular coupling and tissue conductivity producing anisotropy in propagation velocity.

In 8(b) a spiral wave pair is initiated at the fundus, and appears as a periodic source. Such re-entrant waves can be produced by a localised excitation applied at an appropriate time in the wake of a travelling wave and so provide a possible mechanism for premature contractions in a physiologically normal pre-term uterus. Re-entrant waves in two-dimensional media are born in pairs of opposite chirality – and can be characterised by the location of their tips, identified as a phase singularity in physiological recordings, or by the intersection of isolines in computational simulations, and by their frequency. Spiral waves have been induced and recorded in vitro using multiple electrodes in tissue experiments on the internal surface of the guinea pig uterus [Lammers et al., 2008], and may be anticipated to occur in the human pregnant uterus. Their demonstration would require high resolution spatial mapping of uterine activity, and a multiple channel multi-channel SQUID array has been developed to record the spatiotemporal activity in the human uterus [Eswaran et al. 2009] and characterise synchronisation [Ramon et al. 2005]. Multi-channel monitoring of spatiotemporal activity of the uterus is beginning to be used in research and to provide spatiotemporal data of uterine activation patterns before and during early labour. Re-entrant waves are an expected behaviour of an excitable tissue and a computationally predicted behaviour for the uterus. They are a peculiar (pathological) behaviour of a normal (physiological) system, and so are an example of a dynamical disease [May, 1978], and are produced by unusual initial conditions in a normal system. During gestation the uterine myometrium is resting, but can be excitable; localised excitation in the wake of an excitation waves could initiate a pair of spiral waves. A possible mechanism would be an appropriately timed pair of kicks by the foetus, each initiating a localised excitation.   This could provide a mechanism for the onset of premature contractions in physiologically normal myometrium. Topologically, the pregnant myometrium is an excitable, thin walled sphere, with an inexcitable hole (the cervix). A pair of spiral waves in such a system would be expected to meander (the tips move in a cycloid fashion) and drift (a much slower directed motion of the tips); both these are at velocities less than the propagation velocity. These velocities are all increased by any spatial gradients in parameters. Given the geometry, one tip would reach the inexcitable boundary with the cervix, where it would continue rotating around the cervix, leaving one free to in the uterus. This free tip could persist, giving maintained repetitive excitation, or could be extinguished by propagation failure (an accumulation of refractoriness, or a dissipation of excitation), or also drift to the cervix, and extinguish the re-entry. Thus re-entrant excitation could be persistent and pharmacologically intractable (as the cell and tissue parameters are all normal), or self-terminating.

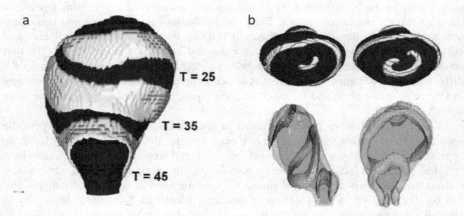

Fig. 8. Surface views of computed excitation pattern, with white/colour excited, blue recovered tissue. The blue and white Fig.s have the wave at different times superimposed. (a) Solitary wave initiated at the fundus, with excitation wave front shown at three successive times T. (b) Spiral pair initiated in the fundus appears as persistent periodic activity, as a site emitting annular waves: the diffusion coefficient is reduced to give a shorter wavelength.

8   Discussion and Conclusions

The mechanisms of human birth are difficult to study, as patient care has a higher priority than research. Computational modeling provides a route for quantitatively linking non-invasive clinical measurements with the results of in vitro cell and tissue studies on biopsies obtained during Caesarean sections. These lead to a view of the control of uterine activity by nonlinear wave processes in excitable media, and to changes in uterine electrophysiology during pregnancy as a trajectory in excitation and coupling parameter space.

During most of pregnancy the uterus is quiescent, has a low excitability, and weak coupling prevents propagation and global contractions. As intercellular coupling increases during gestation propagation can occur, with a slow conduction velocity that gives a short wave length, that could allow re-entrant propagation. Re-entrant waves (an abnormal pattern of activity in a physiologically normal system) could be one mechanism of premature labour, but would be expected to self-terminate unless pinned to some structural heterogeneity. Further increases in coupling gives longer wavelength that could account for the synchronous, global contractions of normal labour.

This quantitative approach is focused on the electrophysiology of the uterine smooth muscle, and views uterine contraction as the primary control target. This neglects the necessity for mechanical changes in cervix (dilation and "ripening"), and the extensive maternal, foetal and placental endocrine interactions that influence uterine myometrial excitability and intercellular coupling, and cervical mechanics.

References

Aslanidi, O., Atia, J., Benson, A.P., can den Berg, H.A., Blamks, A.M., Choi, C., Gilbert, S.H., Goryanain, I., Hayes-Gill, B.R., Holden, A.V., Li, P., Norman, J.E., Shymygol, A., Simpson, N.A.B., Taggart, M.J., Tong, W.C., Zhang, H.: Towards a computational reconstruction of the electrodymamics of premature and full term labour. Prog. Biophys. Mol. Biol. 107, 182–192 (2011)

Bertram, R., Butte, M.J., Kiemel, T., Sherman, A.: Topological and phenomenological classification of bursting oscillations. Bull. Math. Biol. 57, 413–439 (1995)

Bru-Mercier, G., Shmygol, A., Thornton, S., Blanks, A.M.: Spontaneous activity and the complex action potential requires the myometrial network. Reproductive Sciences 15(2), S114A–S115A (2007)

Burdyga, T., Borisova, L., Burdyga, A.T., Wray, S.: Temporal and spatial variations in spontaneous Ca events and mechanical activity in pregnant rat myometrium. Eur. J. Obstet. Gynecol. Reprod. Biol. 144(suppl. 1), S25–S32 (2009)

Bursztyn, L., Eytan, O., Jaffa, A.J., Elad, D.: Mathematical model of excitation-contraction in a uterine smooth muscle cell. Am J. Physiol. 292, C1816–C1829 (2007)

Eswaran, H., Govindan, R.B., Furdea, A., Murphy, P., Lowery, C.L., Priessl, H.T.: Extraction, quantification and characterization of uterine magnetomyographic activity-a proof of concept case study. European J. of Obstet. & Gynecol. Reprod. Biol. 144, 96–100 (2009)

Franciscus, R.: When did the modern human pattern of childbirth arise? New insights from an old Neathertal pelvis. Proc. Nat. Acad. Sci. (USA) 106, 9125–9126 (2009)

Garfield, R.E., Maner, W.L.: Physiology and electrical activity of uterine contractions. Seminars in Cell and Developmental Biology 18, 289–295 (2007)

May, R.: Dynamical diseases. Nature 272, 673–674 (1978)

Inoue, Y., Okabe, K., Soeda, H.: Augmentation and suppression of action potentials by estradiol in the myometrium. Can. J. Physiol. Pharm. 77, 447–453 (1999)

Lammers, W.J.E.P., Mirghani, H., Stephen, B., Dhanasekaran, S., Wahab, A., Al Sultan, M.A.H., Abazer, F.: Patterns of electrical propagation in the intact pregnant guinea pig uterus. Am. J. Physiol. Regul. Integr. Comp. Physiol. 294, R919–R928 (2008)

Norman, J.E., Morris, C., Chalmers, J.: The Effect of Changing Patterns of Obstetric Care in Scotland (1980–2004) on Rates of Preterm Birth and Its Neonatal Consequences: Perinatal Database Study. PLoS Med. 6(9), e1000153 (2009)

Norwitz, E.R., Robinson, J.N., Challis, J.R.G.: The control of labour New England. J. Medicine 341, 66666 (1999)

Partington, H.C., Tonta, M.A., Brennecke, S.P., Coleman, H.A.: Contractile activity, membrane potential, and cytoplasmic calcium in human uterine smooth muscle in the third trimester of pregnancy and during labour. American J. Obsetrics and Gynaecolgy 181, 1145–1151 (1999)

Ramon, C., Preissl, H., Murphy, P., et al.: Synchronization analysis of the uterine magnetic activity during contractions. Bio. Med. Eng. OnLine 4(55) (2005)

Rihana, S., Terrien, J., Geramain, G., Marque, C.: Mathematical modelling of the activity of uterine muscle cells. Med. Biol. Eng. Comput. 47, 665a–675a (2009)

Shmigol, A.V., Eisner, D.A., Wray, S.: Properties of voltage-activated [Ca++]i transients in single smooth muscle cells isolated from pregnant rat uterus. Journal of Physiology 511, 803–811 (1998)

Shmygol, A., Blanks, A.M., Bru-Mercier, G., Gullam, J.E., Thornton, S.: Control of Uterine Ca2+ by Membrane Voltage Toward Understanding the Excitation–Contraction Coupling in Human Myometrium. Ann. N. Y. Acad. Sci. 1101, 97–109 (2007)

Tong, C., Choi, C.Y., Kharche, S., Holden, A.V., Zhang, H., Taggart, M.Y.: A computational model for the ionic currents, Ca2+ dynamics and action potentials underlying contraction of isolated uterine smooth muscle. PLoS ONE 6(4), e18685 (2011)

Winfree, A.T.: Varieties of spiral wave behaviour: An experimentalist's approach to the theory of excitable media. Chaos 1, 303–334 (1991)

Wolf, G.M.J.A., van Leeuwen, M.: Electromyographic observations on the human uterus during labour. Acta Obstet. Gynecology Scand. Suppl. 90, 2–61 (1979)

Evolving Computational Dynamical Systems to Recognise Abnormal Human Motor Function

Michael A. Lones[1], Stephen L. Smith[1], Andy M. Tyrrell[1],
Jane E. Alty[2], and D.R. Stuart Jamieson[2]

[1] Department of Electronics, University of York,
Heslington, York YO10 5DD, UK
{mal503,sls,amt}@ohm.york.ac.uk
[2] Leeds General Infirmary, Leeds, LS1 3EX, UK

Abstract. Artificial biochemical networks (ABNs) are a class of computational automata whose architectures are motivated by the organisation of genetic and metabolic networks. In this work, we investigate whether evolved ABNs can carry out classification when stimulated with time series data collected from human subjects with and without Parkinson's disease. The evolved ABNs have accuracies in the region of 80-90%, significantly higher than the diagnostic accuracies typically found in initial clinical diagnosis. We also show that relatively simple ABNs, comprising only a small number of discrete maps, are able to recognise the abnormal patterns of motor function associated with Parkinson's disease.

1 Introduction

We recently developed a series of computational dynamical systems motivated by the structure and function of biochemical networks [7,8]. When evolved using an evolutionary algorithm, these *artificial biochemical networks* were shown to be competent at solving a diverse range of difficult control problems. In this work, we investigate whether artificial biochemical networks can be used to solve a difficult classification task, by distinguishing between movement time series data collected from Parkinson's disease patients and age-matched controls in a recent clinical study. In particular, we look at the classification accuracy of artificial biochemical networks which are composed of only a few non-linear discrete maps. Discrete maps, such as the logistic map and Chirikov's standard map, model complex real-world processes using simple iterative equations. Their dynamics make them computationally interesting in their own right, and when coupled together they have been shown to carry out difficult computational tasks [1].

2 Materials and Methods

2.1 Data Sets

Parkinson's Disease (PD) is a chronic neuro-degenerative disorder caused by the loss of dopamine-generating cells in the brain. The symptoms of PD are

M.A. Lones et al. (Eds.): IPCAT 2012, LNCS 7223, pp. 177–182, 2012.

highly variable, but all patients develop some form of movement disorder—such as slowing of movement (*bradykinesia*), tremor, rigidity, and impaired balance. Because of its symptomatic diversity, and symptom overlap with other diseases, PD is sometimes difficult to diagnose, with clinical misdiagnosis rates in the region of 25% [2,5].

In a recent clinical study, we collected movement data from 49 PD patients and 41 age-matched controls as they performed a finger tapping task, a standard clinical means of measuring bradykinesia. The subject was asked to tap their thumb and index finger repeatedly for a duration of 30 seconds, using each hand in turn. Movement data was collected using a Polhemus Patriot electromagnetic motion tracking device, whose probes were attached to the subject's thumb and index finger whilst carrying out the task. The movement data was then transformed into time series of displacements between thumb and index finger, and divided into training and test sets in the ratio 2:1. The training data was used for fitness evaluation, and the test set was used to measure classifier generality.

2.2   Classifier Architectures

Artificial biochemical networks (ABNs) are a class of computational automata whose form and function are loosely modelled upon the biochemical networks found within biological organisms. In [7] and [8] we developed various ABN models, and showed how they display rich computational behaviours when coupled to a spectrum of dynamical systems. In this work, we are interested in whether ABNs can perform classification when stimulated with patient movement data. This approach is based on the hypothesis that ABNs can be evolved which will react to the dynamics found within a movement time series, causing them to alter their internal state in an observable manner. The approach is comparable to other uses of computational dynamical systems to perform time series classification, for example recurrent neural networks [3].

In this work, we use two types of ABN: artificial metabolic networks (AMNs) and artificial genetic networks (AGNs). The former are loosely modelled upon metabolic networks, and involve a set of functional elements (termed *enzymes*) manipulating an indexed set of real numbers (*chemical concentrations*) over a period of iterations. In order to mimic the effect of mass conservation in biology, the sum of chemical concentrations are normalised after each iteration. AGNs are a model of genetic regulation, and comprise an indexed array of *genes*, each of which has a real-valued state (*expression level*), a function (*regulatory function*) and a set of inputs (indices of other genes). The AGNs are synchronously updated over a period of iterations, with each gene's expression level at each iteration determined by applying its regulatory function to the expression levels of its input genes. Enzyme and regulatory functions are selected from a set of non-linear discrete maps: the logistic map, Chirikov's standard map, the baker's map, and Arnold's cat map [8]. Between them, these maps display a wide range of ordered and chaotic dynamical behaviours.

Inputs are delivered to the ABNs by setting a chemical concentration or the expression level of a designated input gene. Outputs are read from the final state

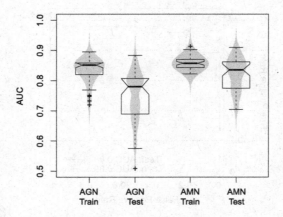

Fig. 1. Diagnostic power of evolved ABNs on both the training and test sets. Notched box plots show summary statistics over 50 runs. Overlapping notches indicate when median values (thick horizontal bars) are not significantly different at the 95% confidence level. Kernel density estimates of underlying distributions are also shown (in grey).

of a designated chemical concentration or gene expression level. A time series is delivered to a network one value at a time, each followed by t_b iterations of the network. Once the whole time series has been delivered, the network is executed for another t_a iterations in order to allow the dynamics to settle. At this point a single output value is read, which is then interpreted as the network's classification of the sequence. The settling parameters, t_b and t_a are both evolved with the network.

2.3   Evolutionary Algorithm

We used a standard generational evolutionary algorithm to evolve ABN-based classifiers. This used tournament selection of size 4, a single elite, a point mutation rate of 6% and uniform crossover with a crossover probability of 15%. Initial solution sizes were made intentionally short, between 2 and 10 genes/enzymes, to encourage parsimony. Each evolutionary run had a population size of 200 and a generation limit of 100. The fitness function was the area under the ROC curve (AUC), which is equivalent to the probability that a network will generate a higher output value for a PD time series than one from a control subject. Its relationship to probability means that AUC is easy to interpret, making it a popular metric in medicine [4].

3   Results

Fig. 1 shows the distribution of training and test scores for the best classifiers from each of 50 evolutionary runs, showing that AUC scores approach 0.9 for the

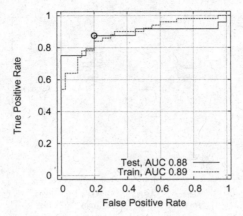

Fig. 2. ROC curves showing the diagnostic accuracy of an evolved classifier across all thresholds. The best trade-off is marked with a circle, showing a sensitivity of 87.5% and a specificity of 80%.

best parameter sets. This corresponds to classification accuracies in the range of 80-90% (see Fig. 2 for an example ROC curve). Whilst lower than the 92-94% accuracy of diagnosis performed by experts in movement disorders, it is considerably higher than the diagnostic accuracies found in non-expert secondary care (75%) and community care (47%), where most patients are first diagnosed [9]. This level of misdiagnosis led to the UK's National Institute of Clinical Excellence (NICE) to recommend that suspected PD patients should only be diagnosed by specialists. However, with this level of diagnostic accuracy, it is feasible that these kinds of classifiers could be used to assist primary care professionals such as general practitioners—especially given the relatively low cost of the equipment and the non-invasive nature of the diagnostic process.

Unlike our earlier diagnostic work [6,10], which used window-based GP classifiers, ABNs have access to both local (e.g. local patterns of acceleration) and global (e.g. spectral characteristics) features of the data, so in principle are able to base their classification on diverse factors. This seems particularly important when processing movement data from Parkinson's patients, where symptom diversity means that individual indicators have poor diagnostic accuracy. For instance, the presence of tremor (measured using spectral analysis of at-rest data collected during the same clinical trial) has a diagnostic accuracy of 63%.

Our results are also interesting from an information processing perspective. Whilst the best classifiers generally make use of several discrete maps, good classification accuracy can be achieved with networks containing only one or two functional elements. Fig. 3 gives an example of this, showing the behaviour of an AMN containing only a single discrete map—Chirikov's standard map operating within a majority chaotic phase. It is surprising that a single chaotic map, in concert with the mass conservation rule of the AMN, can achieve a relatively high classification accuracy (AUC=0.84). Furthermore, this capacity does not

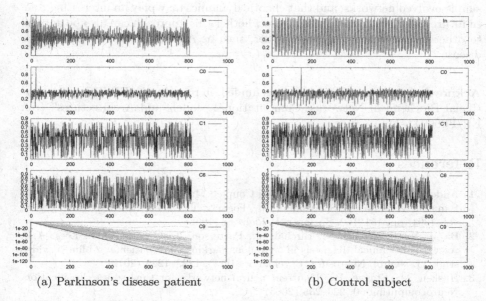

Fig. 3. Example of an evolved AMN processing movement data (In) from subjects with and without Parkinson's. The lower four plots in each case show how chemical concentrations change as the input sequence is processed. C9 is the designated output chemical, whose final concentration is interpreted as the classifier's output. For comparison, grey lines show the C9 time series for all members of the test set.

appear to be linked to the choice of discrete map, since solutions with similar classification accuracies were evolved which contained each of the discrete maps available to the evolutionary algorithm.

Given the gap between ABNs and the biological structures which they are motivated by, it is hard to say whether these results give any insight into the nature of biochemical information processing. However, it does show that relatively simple (from an implementation perspective) non-linear processes can process signals produced by a relatively complex biological process. These non-linear processes, in turn, occur in many naturally occurring systems, so it doesn't seem unreasonable that they could occur in the biochemical networks present within cells and tissues.

4   Conclusions

In this paper, we have shown that artificial biochemical networks can be used to recognise abnormal motor function associated with Parkinson's disease. The evolved classifiers perform an objective diagnosis based upon data collected from simple movement tasks, and have accuracies comparable to trained clinicians. Analysis of the classifiers suggests that diagnosis can be performed by relatively simple evolved networks, and that chaotic dynamics may play an interesting role. In future work, we hope to investigate whether this approach can also be applied to other forms of neurological disorder, such as Alzheimer's and Huntington's disease.

Acknowledgements. This work was funded, in part, by the EPSRC grant "Artificial Biochemical Networks: Computational Models and Architectures" (ref: EP/F060041/1).

References

1. Andersson, C., Nordahl, M.: Evolving Coupled Map Lattices for Computation. In: Banzhaf, W., Poli, R., Schoenauer, M., Fogarty, T.C. (eds.) EuroGP 1998. LNCS, vol. 1391, pp. 151–162. Springer, Heidelberg (1998)
2. Bajaj, N.P.S., Gontu, V., Birchall, J., Patterson, J., Grosset, D.G., Lees, A.J.: Accuracy of clinical diagnosis in tremulous parkinsonian patients: a blinded video study. J. Neurol. Neurosurg. Psychiatry 81(11), 1223–1228 (2010)
3. Hüsken, M., Stagge, P.: Recurrent neural networks for time series classification. Neurocomputing 50, 223–235 (2003)
4. Kraemer, H.C., Morgan, G.A., Leech, N.L., Gliner, J.A., Vaske, J.J., Harmon, R.J.: Measures of clinical significance. J. Am. Acad. Child. Adolesc. Psychiatry 42(12), 1524–1529 (2003)
5. Levine, C.B., Fahrbach, K.R., Siderowf, A.D., Estok, R.P., Ludensky, V.M., Ross, S.D.: Diagnosis and treatment of Parkinson's disease: a systematic review of the literature. Evid. Rep. Technol. Assess. (Summ.) (57), 1–4 (2003)
6. Lones, M.A., Smith, S.L.: Objective assessment of visuo-spatial ability using implicit context representation cartesian genetic programming. In: Smith, S.L., Cagnoni, S. (eds.) Genetic and Evolutionary Computation: Medical Applications, John Wiley & Sons, Chichester (2010)
7. Lones, M.A., Tyrrell, A.M., Stepney, S., Caves, L.S.D.: Controlling Complex Dynamics with Artificial Biochemical Networks. In: Esparcia-Alcázar, A.I., Ekárt, A., Silva, S., Dignum, S., Uyar, A.Ş. (eds.) EuroGP 2010. LNCS, vol. 6021, pp. 159–170. Springer, Heidelberg (2010)
8. Lones, M.A., Tyrrell, A.M., Stepney, S., Caves, L.S.D.: Controlling legged robots with coupled artificial biochemical networks. In: Lenaerts, T., et al. (eds.) Proc. 11th European Conference on the Synthesis and Simulation of Living Systems, Advances in Artificial Life, ECAL 2011, pp. 465–472. MIT Press (August 2011)
9. National Institute for Health and Clinical Excellence: Parkinson's disease: diagnosis and management in primary and secondary care. Royal College of Physicians (2006), http://www.nice.org.uk/CG035
10. Smith, S.L., Gaughan, P., Halliday, D.M., Ju, Q., Aly, N.M., Playfer, J.R.: Diagnosis of Parkinson's disease using evolutionary algorithms. Genetic Programming and Evolvable Machines 8(4), 433–447 (2007)

Heat-Maps and Visualization
for Heterogeneous Biomedical Data
Based on Information Distance Geometry

Esther Loeliger[1], Chrystopher L. Nehaniv[1,2], and Alastair J. Munro[3]

[1] Algorithms Research Group
University of Hertfordshire
United Kingdom
e.loeliger@herts.ac.uk
[2] Royal Society Wolfson BioComputation Research Lab
University of Hertfordshire
United Kingdom
[3] Department of Population Health Sciences,
University of Dundee,
United Kingdom

Abstract. Systems biology is very much concerned with gaining an overview of what is happening in complex systems, such as in biomedical data sets, for which we need good global visualization tools. This research uses a method based on information distance geometry to create visualizations analogous to heat-maps of prognostic and diagnostic variables. It illustrates the advantages of an informationally self-structuring approach to the understanding of biomedical data.

Keywords: heat-maps, data visualization, information distance, heterogeneous biomedical data, systems biology.

We present a method of analysing and visualizing biomedical data that uses raw, uninterpreted data points of varying modalities as input. The data is a highly abstracted data set derived from a population of cancer patients. Rather than pre-select indicators that seem likely to influence clinical outcomes, each one of more than eighty prognostic and diagnostic variables is analysed in information theoretical terms and related to outcomes such as the site of the primary tumour and survival. The goal is to bring as little baggage as possible to the task of finding out which variables predict the site of the tumour, and which variables influence survival. This approach is based on mathematical formulations, not judgements grounded in prior experience or theory.

One key argument in favour of using information distance is that it is a mathematical metric, i.e. it gives distances in a geometric space, which satisfy triangle inequality, symmetry and positive definiteness. It 'captures general relationships between sensors and not just linear relationships'[5]. It allows one to find the appropriate dimensionality of the representation of the data. No other method described below is a metric in this sense.

M.A. Lones et al. (Eds.): IPCAT 2012, LNCS 7223, pp. 183–187, 2012.
© Springer-Verlag Berlin Heidelberg 2012

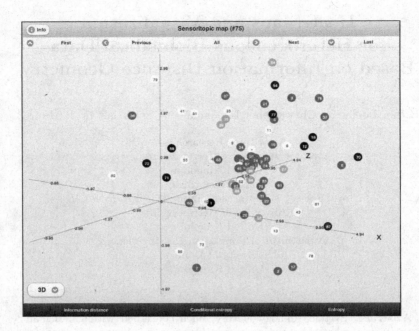

Fig. 1. The 3D map shows a node for each column in the biomedical data. These nodes have been arranged in a 3D space in such a way that their informational distance to each other is reflected in the straight-line distance between them in the figure. The information distance measures how much information is in the variables that they don't tell about each other. This distance is measured in bits (in the sense of Shannon information theory). So if two variables are very close then one can be used to predict the other one, even if the means of predicting is not at all linear. Colours on these nodes correspond to particular patients (rows in your data). The colour of a node indicates the value of the corresponding variable for that patient : red indicates low, black middle and green high values. White nodes represent missing values.

Slonim et al. have argued that clustering by similarity or prototype often requires one to 'specify the similarity measure in advance', and also that this approach can lend itself to arbitrary choices[6]. Their proposed method centres instead on 'information as a Similarity Measure'. The algorithm uses mutual information to highlight meaningful data clusters in fields as disparate as gene expression, stock prices and movie ratings. The absence of a prior selection criterion leads to the detection of patterns involving data that had previously been excluded as 'stereotypical'[6]. The wide range of inputs demonstrates that Slonim et al.'s approach works well across domain boundaries.

Olsson, Nehaniv and Polani apply information distance – Crutchfield's 'information metric'[1] – to an analysis of unstructured sensor data received by a robot, building sensoritopic maps for a range of motor settings[4]. Here again, the emphasis is on unfiltered input data and the tools of information theory,

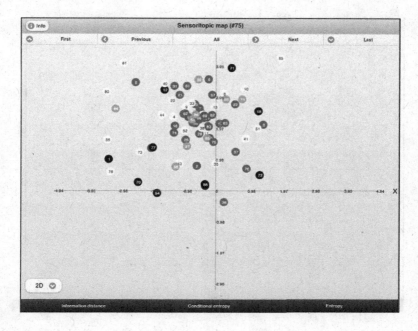

Fig. 2. The 2D map shows a node for each column in the biomedical data; for more details see caption fig. 1

making the robot resilient to changes in its environment and sensory apparatus much as Slonim et al.'s algorithm makes the leap from gene expression to stock prices.

Olsson, Nehaniv and Polani compare information distance to five alternative distance measures including Kullback-Leibler, Hellinger and Jensen-Shannon. Having examined sensoritopic maps based on all six methods, they conclude that information distance is 'the only successful method, outperforming all the other distance measures'[5]. They show information distance, when combined with entropy maximization, allows for the best integration of variables coming from different modalities.

Current heat-map methods for gene expression are generally limited to a pre-specified, usually one-dimensional ordering of the variables, with the ordering being imposed by some clustering method. In the case of Slonim et al., this method is mutual information. Information distance allows one instead to use the natural dimensionality in the data themselves, and does not presume a linear ordering. It can recover geometric structure in which to place the information sources automatically, including dimensionality of the representation, but also including the shape in terms of where the variables are most naturally located with respect to one another[3].

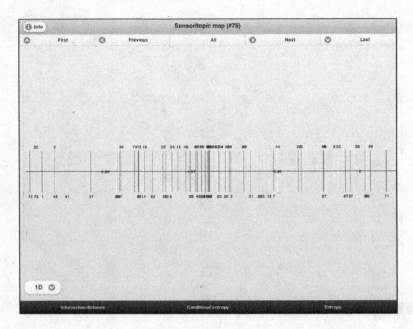

Fig. 3. The 1D map shows a line for each column in the biomedical data; for more details see caption fig. 1

Nehaniv et al. have defined the 'utility measure' of a given sensory channel in terms of its ability to 'carry only relevant information and to discard the rest'[2]. The experiments of Olsson, Nehaniv and Polani have demonstrated that information distance is both highly versatile with regard to inputs spanning multiple domains and particularly likely to identify relevant relationships.

This research uses information distance geometry to create visualizations analogous to heat-maps of prognostic and diagnostic variables in a large collection of heterogeneous biomedical data. This approach should help us find an appropriate dimensionality and geometry for the representation of the data. Once the geometry is set up, the implementation will show the values of particular patients in colours red to green and highlight the variables identified as the most important or relevant.

Users benefit from appropriate representations to manipulate and understand large unstructured data sets, which is a common problem in the analysis of biological data such as gene expression arrays. This work combines our method with others and illustrates advantages of an informationally self-structuring approach to the understanding of biomedical data.

References

1. Crutchfield, J.P.: Information and its metric. In: Nonlinear Structures in Physical Systems Pattern Formation, Chaos and Waves, pp. 119–130. Springer, Heidelberg (1990)

2. Nehaniv, C.L., Polani, D., Dautenhahn, K., te Boekhorst, R., Canamero, L.: Meaningful information, sensor evolution, and the temporal horizon of embodied organisms. In: Artifical Life VIII, pp. 345–349. MIT Press (2002)
3. Olsson, L.: Information Self-structuring for Developmental Robotics: Organization, Adaptation, and Integration. PhD thesis, University of Hertfordshire, School of Computer Science, United Kingdom (2006)
4. Olsson, L., Nehaniv, C.L., Polani, D.: Discovering motion flow by temporal-informational correlations in sensors. In: Fifth International Workshop on Epigenetic Robotics: Modeling Cognitive Development in Robotic Systems, pp. 117–120 (2005)
5. Olsson, L., Nehaniv, C.L., Polani, D.: Measuring informational distances between sensors and sensor integration. In: Rocha, L.M., Yaeger, L.S., Bedau, M.A., Floreano, D., Goldstone, R.L., Vespignani, A. (eds.) Artificial Life X: Proceedings of the Tenth International Conference on the Simulation and Synthesis of Living Systems, pp. 316–322. MIT Press/Bradford Books (2006)
6. Slonim, N., Atwal, G.S., Tkacik, G., Bialek, W.: Information-based clustering. Proc. Natl. Acad. Sci. U S A 102(51), 18297–18302 (2005)

Bio-inspired Information Processing Applied to Engineering Systems

Cristina Costa Santini

Information Technology Department,
College of Computer and Information Sciences,
King Saud University, P.O.Box 22452, Riyadh 11495, Kingdom of Saudi Arabia

Abstract. Over the course of billions of years, under evolutionary pressure, Nature has evolved solutions to various problems. As our ability to understand the biological mechanisms that are intrinsic in these solutions continues to improve, we have the opportunity to apply this knowledge when solving our challenging problems, in fields such as medicine and the environment. This paper discusses an approach, in which biological systems are investigated as information processing systems, and the understanding of how these systems process information is then applied to engineering systems. Two examples are presented. The first one discusses how the heart's fault-tolerant information processing can be implemented in an electronic system. The second example discusses a cellular biochemical reaction network and how its property of robustness can be implemented in a chemical system. Finally, three different applications, in which this approach is already being applied with promising results, are briefly reviewed.

Keywords: information processing, bio-inspired engineering systems.

1 Introduction

This paper discusses an approach, in which molecular, cellular, and organism-level biological systems are investigated as information processing systems, and the understanding of how these systems process information is then applied to engineering systems, as opposed to algorithms [1, 2]. Algorithms such as cellular automata, neural networks, and evolutionary computation use Nature as inspiration and have been proposed in the 1950s [3–5]. Other bio-inspired algorithms, namely swarm intelligence, artificial immune systems, membrane computing and amorphous computing, are more recent [6–9].

The focus here is to discuss the engineering of systems that implement Nature's computational paradigms, mimicking biological systems' problem solving strategies. This approach has the potential to contribute to the solution of challenging problems, in fields such as environment and medicine.

Definitions of information have already been given by different people in different contexts [10]. Claude Shannon proposed that information implies surprise, being closely related with uncertainty, and that it carries no meaning [11].

M.A. Lones et al. (Eds.): IPCAT 2012, LNCS 7223, pp. 188–199, 2012.

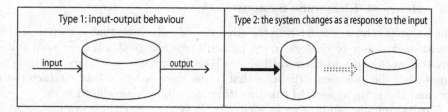

Type 1: input-output behaviour	Type 2: the system changes as a response to the input

Fig. 1. Abstract representation of an information processing system. Two different types of information processing systems are considered. Type 1: given an input, the system will produce an output that is a function of the input, with no changes to the system; Type 2: the system itself changes as a response to the input.

The Russian Kolmogorov proposed, at around the same time as the American Chaitin, what nowadays is referred to as the Kolmogorov-Chaitin algorithmic information theory [12, 13]. According to this theory, the size of the algorithm that is needed to generate a given string measures how much information it contains.

The two definitions given above are both abstract definitions. In this paper, two different types of information processing systems are considered (Fig. 1). A system of type 1 is a system that, given an input, will produce an output that is a function of the input, whereas a system of type 2 is a system that changes itself as a response to the input. Throughout the paper, this abstract representation of an information processing system is used. This same representation will illustrate the different systems discussed. Note that in these systems the terms input and output are used very broadly and will hold diverse meanings.

This paper is organised as follows. Section 2 presents examples where biological systems have been investigated as information processing systems and the gained knowledge has been used to design engineering systems. Section 3 reviews promising application areas and Section 4 concludes the paper.

2 Biological Investigation and Artificial Implementation

In order to apply bio-inspired information processing to engineering systems, the biological system has to be investigated and its information processing aspect characterised. Then the context of the biological system and the scope of the artificial implementation have to be carefully considered.

In this section, two systems will be described, exclusively from the information processing perspective. The first one discusses how the heart processes information in a fault-tolerant and robust way and shows how this information processing paradigm can be implemented in an electronic system. The second example discusses a cellular biochemical reaction network and how its property of robustness can be implemented in a chemical system.

In both examples it is their robust characteristic that is being investigated and mimicked in the artificial systems. Nevertheless, the diverse nature of these two natural systems show robustness related to different aspects of the system.

2.1 Heart ⇒ Electronic System

The heart is robust, considering the diversity of influences that act upon it [14] but also adaptable to changes in the person's psychological and emotional state, physical workload and chemical balance [15]. In previous work [16], we investigated what the biological principles that confer the heart its robust characteristic are, and the main aspects of this investigation will be reproduced here.

Cardiac muscle can be seen as an excitable system. Excitable systems exhibit rhythmic properties or limit cycles. They are as diverse as chemical reactions and hepatocytes and share some common characteristics, namely self-organisation and robustness. This is due to the fact that in these systems the overall behaviour, i.e. their functionality is not a result of the function of single elements but a consequence of the interaction of these elements.

The qualitative behaviour of cardiac cells can be modeled by the FitzHugh-Nagumo (FHN) model [17], which is a generic model for excitable media. FitzHugh called his simplified model the Bon Hoeffer-van der Pol model (BVP) and derived it as a simplification of the Hodgkin-Huxley [18] equations. The model is able to reproduce many qualitative characteristics of electrical impulses along nerve and cardiac fibres and is described by the following pair of differential equations:

$$\frac{dx}{dt} = c(x + y - \frac{x^3}{3} + z) \tag{1}$$

$$\frac{dy}{dt} = -\frac{1}{c}(x - a + by), \tag{2}$$

where a, b and c are constants satisfying the relations

$$1 - \frac{2b}{3} < a < 1, 0 < b < 1, b < c^2 \tag{3}$$

and z is stimulating current. The coordinate x shares the properties of both membrane potential and excitability, while y is responsible for accommodation and refractoriness.

This model is used to simulate the qualitative behaviour of two types of cardiac cells: the SAN (sinoatrial node) cells, which are autonomous oscillators and the AM (atrial myocardial) cells, which are excitatory. We simulate a grid of 9 x 9 cardiac cells, with the cells in the centre representing the SAN cells and the other cells representing the AM cells. In the simulation, as in the heart, the SAN cells are not identical. When independent, i.e. not coupled, they fire at different frequencies. When coupled through the diffusion of membrane potential (x in the model), the SAN cells are phase locked, and the AM cells propagate the stimulus.

Results show that synchronisation between the SAN cells and consequent propagation to the AM cells only happens for a certain level of diffusion coefficient. Below this level, the SAN cells oscillate with different frequencies and there is no potential being propagated through the AM cells. The robust functionality of the heart, whose function is to beat and pump blood, depends on the

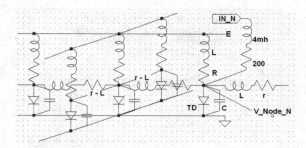

Fig. 2. Circuit Architecture. Each element or 'circuit node' comprises of a tunnel diode (TD), a capacitor (C), an inductor (L) and a resistor (R). These nodes are connected through a resistor (r) and an inductor (L) with zero flux boundary conditions at the edges [16].

synchronisation between the SAN cells. In this synchronised state, the cardiac tissue can tolerate a certain level of damage without the function of the heart being compromised [16].

This robust characteristic, exhibited by the heart, is desirable in artificial systems. In electronic systems for example, components can function differently over time or even fail. We therefore investigated how self-organisation could be harnessed in an artificial system [16]. An electronic oscillator was used as the basic element for this electronic circuit. A circuit composed of 81 of these basic oscillators arranged in a 9 x 9 grid was simulated (Fig. 2).

These electronic oscillators are not identical; they oscillate at different frequencies. Results show that when the resistive coupling between these oscillators is low enough, i.e. when there is enough interaction between these elements, they synchronise. In this synchronised state they exhibit robust, fault-tolerant behaviour. If, for example, an electronic oscillator is removed from the grid, the synchronisation between the other elements of the circuit is maintained. This same conclusion was arrived at after further investigations, where a bigger grid with 300 oscillators and two different coupling configurations were experimentally investigated [19]. Fig. 3 represents the heart and the electronic system as robust information processing systems.

This example discussed the heart, an organ-level biological system. It is the interaction between the elements of this system, namely the cardiac cells, what accounts for its emergent synchronised behaviour. The next section will discuss a molecular level system, a cellular regulatory network.

2.2 Cellular Regulatory Networks ⇒ Chemical System

How does a cell process information in a robust way? Can this computing paradigm be implemented in an artificial system?

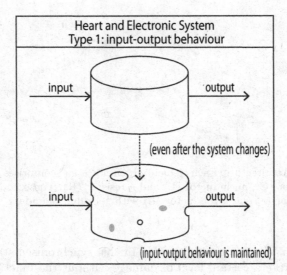

Fig. 3. Abstract representation of the heart and the electronic system as information processing systems. For a certain input (diffusion coefficient d between the cells in the case of the heart and coupling resistance r between the electronic oscillators of the grid) the system is synchronised. This output is robust to changes or faults in the system's components (cardiac tissue and electronic oscillators).

Being able to cope with variations in the number of copies of a gene network in a cell, or network dosage, is crucial for the cell's gene networks function. By using the yeast galactose network as a model, Acar et al. [20] found that network activity was robust to the change in network dosage. After methodological investigations the authors concluded that, in general, two network components, one positive and one negative regulator, is the minimal requirement for network dosage invariance.

Robustness is a desirable characteristic in implementations of molecular systems. We therefore investigated how this biological information processing mechanism could be implemented in a molecular system. Here, simulation results are shown for a general chemical reaction system. This general system is based on the type-1 incoherent feed-forward loop (I1-FFL) network motif.

Network motifs are small, recurrent regulatory subnetworks, classifiable in terms of function, architecture, dynamics, or biochemical process [21–23]. In these networks, the nodes are genes and the edges represent transcriptional regulation of one gene by the protein product of another gene. In the I1-FFL (Fig. 4a), the direct path is positive and the indirect path is negative, i.e. X activates Z and Y, an inhibitor of Z. Experimental and computational approaches have shown that the I1-FFL is responsible for functions such as pulse generation [24], fold-change detection [25] and amplitude filtering [26]. Fig. 4b shows an example of pulse generation using the I1-FFL network motif.

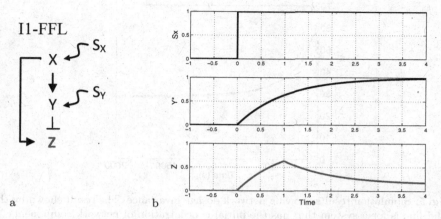

Fig. 4. Type-1 incoherent feed-forward loop (I1-FFL) network motif. **a** Arrows denote activation and $\perp$ denotes repression. S_x and S_y are input signals that activate transcription of genes X and Y. **b** Pulse-generation dynamics of the I1-FFL following an ON step of S_x in the presence of S_y. After signal Sx activates gene X, the product of X turns on Z but also its repressor Y. Active Z accumulates until Y* levels reaches the repression threshold for the Z promoter. As a result, Z production decreases and its concentration drops, resulting in pulse-like dynamics [27, 28].

The topology of the I1-FFL network motif [27] has two components, one positive and one negative regulator, and thus fits the description of the minimal requirement for a network whose activity is robust to changes in network dosage [20]. For this reason, the general chemical reaction system discussed herein is based on the I1-FFL network motif, and is modelled by the following ordinary differential equations:

$$X + S_x \xrightarrow{k_1} X^* \tag{4}$$

$$X^* + Z \xrightarrow{k_2} Z^* \tag{5}$$

$$Y + S_y \xrightarrow{k_3} Y^* \tag{6}$$

$$X^* + Y^* \xrightarrow{k_4} Inh \tag{7}$$

$$Z^* + Inh \xrightarrow{k_5} InhZ \tag{8}$$

In this general system, (*) represents the active form of the component. The behaviour of the proposed general chemical reaction system was investigated through chemical kinetics simulations. Fig. 5 shows the results for two simulations: the dotted line shows the simulation of a network with twice the concentration of network components than the network shown by the solid line. This result shows that the system's output (Z*) is robust to variation in the concentration of system components. This robustness is illustrated in Fig. 6 through the common abstract representation of an information processing system.

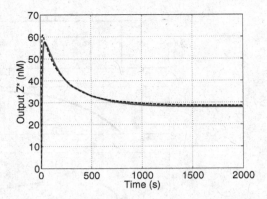

Fig. 5. Simulation results showing network dosage invariance. The result shown by the solid line is for a system that has the initial concentration of network components X, Y and Z and of input S_y set to 1 mM whereas the result shown by the dotted line corresponds to a network that has the initial concentration of network components X, Y and Z and of input S_y set to 2 mM. Both networks have the input stimulus $S_x = 100$ nM. All rate constants k_1 to k_5 are set to $10^5 M^{-1} s^{-1}$ and, for this general system, we assume that the reactions are irreversible. This system could be implemented with, for example, biomolecular species, such as DNA [29]. For this reason, the concentrations and rate constants used are typical values for a DNA based system [30, 31]. (Simulations were performed in Matlab [32] using the ODE solver ode15s.)

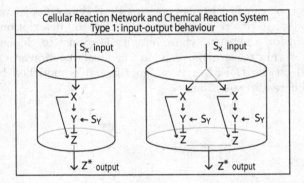

Fig. 6. The activity of the system is robust to change in the number of system components. X is the node that receives the input and Z is the node that transmits the output. A certain input stimulus S_x will cause a certain level of output activation Z* that is invariant to the concentration of network components in the reaction network. The figure illustrates the fact that, even though the system on the right has twice the number of components than the system on the left, the output of both systems is the same.

2.3   Discussion

The examples described in this section show bio-inspired information processing being applied in the engineering of systems that then exhibit robustness to faults and to changes in the system.

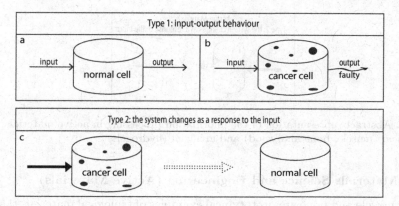

Fig. 7. Abstract representation of a cell. The input to both normal (**a**) and cancer (**b**) cell is the information from outside and from within the cell, and the output, in both types of cells, determine the cell's survival and reproduction. The difference in the cancer cell when compared to the normal cell is that its output is faulty. (**c**) Would it be possible to interact with the molecular signalling network of the cancer cell, in order to therapeutically modulate it, so that it processes information like a normal cell?

Nature had billions of years to evolve highly complex biological systems' mechanisms. With the improved understanding of these mechanisms, we have the opportunity to try mimicking their information processing paradigms when solving the challenging problems facing humanity. The next section discusses three such challenging problems from the information processing perspective.

3   Application Areas

3.1   Medicine (Cancer)

Cells are information processing systems. Molecular signalling networks process information from within and from outside the cell and their responses will determine the cell's survival and reproduction. An important question is therefore: How is information being processed in a cancer cell and how is it different from the way information is processed in a normal cell? The difference is that the responses given by a cancer cell are faulty: they determine the cell's proliferation when they should be quiescent and the cell's survival when they should die [33, 34].

Following the proposed abstract representation for an information processing system, Fig. 7a represents a normal cell and Fig. 7b represents a cancer cell. Fig. 7c illustrates the question concerning if it would be possible to interact with the molecular signalling network of a cancer cell, so that it processes information like a normal cell.

Suggestions of how this could be done have already been discussed. Recently, intracellular mechanisms that can detect a diseased state and deliver the adequate therapy have been proposed [35, 36].

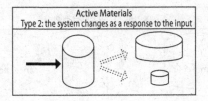

Fig. 8. Abstract representation of the information processing in active materials, both biological (muscle, bone and wood) and artificial (hydrogels)

3.2 Materials Science and Engineering (Active Materials)

Trees, muscle and bone are prototypical examples of biological materials that are responsive to their environment [37, 38]. Muscle can shrink (cell death) or grow (hyperplasia) as a result of stress. The growth of trees respond to environmental conditions such as wind, sun and load. Bones can shrink or grow in response to external load. Fig. 8 shows the abstract representation of the information processing of these biological materials, where the input causes the material to adapt. This adaptation happens on different length scales, from the nanoscopic to the macroscopic scale.

A responsive artificial system that can be represented by this same abstract representation (Fig. 8) is presented in [39], where Sidorenko et al. propose a hybrid architecture, integrating a hydrogel with an array of rigid structures. The hydrogel is responsive to humidity: when humidity increases the hydrogel changes to its swollen state and when humidity decreases the hydrogel changes to its contracted state, exposing the array of rigid structures. The authors suggest that these hybrid architectures may lead to a variety of applications, including microfluidics, as these structures can direct the flow and particle transport (analogous to the function of cilia in organisms).

3.3 Environment (Articial Photosynthesis)

There is eminent need for alternative, renewable and clean sources of energy. The potential of solar energy is an opportunity that should be addressed with urgency [40].

Natural photosynthesis, as Fig. 9a illustrates, converts water, light and carbon dioxide into oxygen and sugars. This overall reaction is the result of several reactions. Amongst these is the water-splitting reaction of photosystem II (PSII), an enzyme found in plants, algae and cyanobacteria. This reaction holds the greatest promise for developing new technologies for converting solar radiation into usable energy, particularly in generating hydrogen [40].

Artificial photosynthesis currently uses two approaches: photo electrochemical and dye-sensitized solar cells. Recently, breakthroughs in both methods have occurred [41, 42]. Both photosynthetic processes convert sunlight into energy, as natural photosynthesis does. However, it is only the photo electrochemical

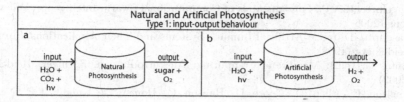

Fig. 9. Abstract information processing representation of **a** natural photosynthesis and **b** artificial photosynthesis

approach that mimics Nature, and in the presence of light, splits water into oxygen and hydrogen (Fig. 9b). For this reason this process is sometimes referred to as the 'artificial leaf'.

4   Conclusion

In this paper, an approach is discussed in which biological systems are investigated as information processing systems, and the understanding of how these systems process information is then applied to engineering systems. Two examples where this approach has been followed are presented and promising applications are reviewed, in the areas of medicine, materials science and engineering and the environment.

This approach is based on interdisciplinarity and can be seen as part of the wider field of biomimicry [43], and as the latter, is motivated by the need to look into Nature for novel solutions to diverse complex problems.

Finally, because bio-inspired engineered solutions are likely to be designed to be integrated with their systems, be it a human being or the environment, it is expected that they will represent a step towards a sustainable world.

Acknowledgments. The author would like to thank Prof. Andy Tyrrell for comments on the manuscript. This study was partially funded by the College of Computer and Information Sciences of King Saud University.

References

1. Kari, L., Rozenberg, G.: The many facets of natural computing. Communications of the ACM 51(10), 72–83 (2008)
2. Navlakha, S., Bar-Joseph, Z.: Algorithms in nature: the convergence of systems biology and computational thinking. Molecular Systems Biology 7(546), 1–11 (2011)
3. Von Neumann, J.: Theory of Self-Reproducing Automata. In: Burks, A.W. (ed.), vol. 21(100). University of Illinois Press (1966)
4. Arbib, M.A.: The Handbook of Brain Theory and Neural Networks. MIT Press (1995)
5. Bäck, T., Fogel, D., Michalewicz, Z.: Handbook of evolutionary computation. Oxford Univ. Press (1997)

6. Engelbrecht, A.P.: Fundamentals of Computational Swarm Intelligence. Wiley and Sons (2006)
7. Dasgupta, D. (ed.): Artificial Immune Systems and Their Applications. Springer, Heidelberg (1998)
8. Paun, G. (ed.): Membrane Computing: An Introduction. Springer, Heidelberg (2002)
9. Abelson, H., Allen, D., Coore, D., Hanson, C., Homsy, G., Knight, T.F., Nagpal, R., Rauch, E., Sussman, G.J., Weiss, R.: Amorphous computing. Communications of the ACM 43(5), 74–82 (2000)
10. Gleick, J.: The Information, 1st edn. Fourth State (2011)
11. Shannon, C.: A mathematical theory of communication. The Bell System Technical Journal 27, 623–656 (1948)
12. Kolmogorov, A.N.: Three approaches to the quantitative definition of information. Problems of Information Transmission 1(1), 1–7 (1965)
13. Chaitin, G.J.: On the length of programs for computing binary sequences. Journal of the ACM 13, 547–569 (1966)
14. Sole, R., Goodwin, B.: Signs of Life: How Complexity Pervades Biology. Basic Books (2002)
15. Winfree, A.T.: When Time Breaks Down: Three-Dimensional Dynamics of Electrochemical Waves and Cardiac Arrhythmias. Princeton University Press (1987)
16. Santini, C.C., Tyrrell, A.: Understanding and harnessing self-organization. In: Proceedings of the 2007 IEEE Symposium on Artificial Life, pp. 192–198. IEEE Computational Intelligence Society (2007)
17. Fitzhugh, R.: Impulses and physiological states in theoretical models of nerve membrane. Biophysical Journal 1, 445–466 (1961)
18. Hodgkin, A.L., Huxley, A.F.: A quantitative description of membrane current and its application to conduction and excitation in nerve. Journal of Physiology 117, 500–544 (1952)
19. Santini, C.C., Tyrrell, A.M.: Investigating the properties of self-organisation and synchronisation in electronic systems. IEEE Transactions on NanoBioscience 8, 237–251 (2009)
20. Acar, M., Pando, B.F., Arnold, F.H., Elowitz, M.B., van Oudenaarden, A.: A general mechanism for network-dosage compensation in gene circuits. Science 329(5999), 1656–1660 (2010)
21. Wolf, D.M., Arkin, A.P.: Motifs, modules and games in bacteria. Current Opinion in Microbiology 6(2), 125–134 (2003)
22. Milo, R., Shen-Orr, S., Itzkovitz, S., Kashtan, N., Chklovskii, D., Alon, U.: Network motifs: simple building blocks of complex networks. Science 298(5594), 824–827 (2002)
23. Shen-Orr, S.S., Milo, R., Mangan, S., Alon, U.: Network motifs in the transcriptional regulation network of escherichia coli. Nature Genetics 31(1), 64–68 (2002)
24. Basu, S., Mehreja, R., Thiberge, S., Chen, M.-T., Weiss, R.: Spatiotemporal control of gene expression with pulse-generating networks. Proceedings of the National Academy of Sciences of the United States of America 101(17), 6355–6360 (2004)
25. Goentoroa, L., Shoval, O., Kirschner, M., Alon, U.: The incoherent feedforward loop can provide fold-change detection in gene regulation. Molecular Cell 36, 894–899 (2009)
26. Kaplan, S., Bren, A., Dekel, E., Alon, U.: The incoherent feed-forward loop can generate non-monotonic input functions for genes. Molecular Systems Biology 4, 203 (2008)

27. Alon, U.: Network motifs: theory and experimental approaches. Nature Reviews Genetics 8(6), 450–461 (2007)
28. Alon, U.: An Introduction to Systems Biology Design Principles of Biological Circuits, 1st edn. Mathematical and Computational Biology Series. Chapman and Hall/CRC (2007)
29. Turberfield, A.: DNA as an engineering material. Physics World, 43–46 (March 2003)
30. Yurke, B., Mills Jr., A.P.: Using DNA to power nanostructures. Genetic Programming and Evolvable Machines 4(2), 111–122 (2003)
31. Green, S.J., Lubrich, D., Turberfield, A.J.: DNA hairpins: Fuel for autonomous DNA devices. Biophysical Journal 91(8), 2966–2975 (2006)
32. MATLAB, version 7.10.0 (R2010a). The MathWorks Inc., Natick, Massachusetts (2010)
33. Hanahan, D., Weinberg, R.: Hallmarks of cancer: The next generation. Cell 144(5), 646–674 (2011)
34. Tyson, J.J., Baumann, W.T., Chen, C., Verdugo, A., Tavassoly, I., Wang, Y., Weiner, L.M., Clarke, R.: Dynamic modelling of oestrogen signalling and cell fate in breast cancer cells. Nature Reviews Cancer 11(7), 523–532 (2011)
35. Xie, Z., Wroblewska, L., Prochazka, L., Weiss, R., Benenson, Y.: Multi-input RNAi-based logic circuit for identification of specific cancer cells. Science 333(6047), 1307–1311 (2011)
36. Venkataraman, S., Dirks, R.M., Ueda, C.T., Pierce, N.A.: Selective cell death mediated by small conditional RNAs. Proceedings of the National Academy of Sciences of the United States of America 107(39), 16777–16782 (2010)
37. Weinkamer, R., Fratzl, P.: Mechanical adaptation of biological materials — the examples of bone and wood. Materials Science and Engineering C 31(6), 1164–1173 (2010)
38. Charg, S.B.P., Rudnicki, M.A.: Cellular and molecular regulation of muscle regeneration. Physiological Reviews 84(1), 209–238 (2004)
39. Sidorenko, A., Krupenkin, T., Taylor, A., Fratzl, P., Aizenberg, J.: Reversible switching of hydrogel-actuated nanostructures into complex micropatterns. Science 315(5811), 487–490 (2007)
40. Barber, J.: Biological solar energy. Philosophical Transactions of the Royal Society - Series A: Mathematical, Physical and Engineering Sciences 365(1853), 1007–1023 (2007)
41. Yella, A., Lee, H.W., Tsao, H.N., Yi, C., Chandiran, A.K., Nazeeruddin, M.K., Diau, E.W.G., Yeh, C.Y., Zakeeruddin, S.M., Gratzel, M.: Porphyrin-sensitized solar cells with cobalt (II/III)-based redox electrolyte exceed 12 percent efficiency. Science 334(6056), 629–634 (2011)
42. Reece, S.Y., Hamel, J.A., Sung, K., Jarvi, T.D., Esswein, A.J., Pijpers, J.J.H., Nocera, D.G.: Wireless solar water splitting using silicon-based semiconductors and earth-abundant catalysts. Science (September 2011)
43. Benyus, J.M.: Biomimicry: Innovation Inspired by Nature. Harper Perennial (1997/2002)

Understanding the Regulation of Predatory and Anti-prey Behaviours for an Artificial Organism

Maizura Mokhtar

School for Computing, Engineering and Physical Sciences,
University of Central Lancashire (UCLan), Preston, PR1 2HE, UK
MMokhtar@uclan.ac.uk

Abstract. An organism's behaviour can be categorised as being either predatory or anti-prey. Predatory behaviours are behaviours that try to improve the life of an organism. Anti-prey behaviours are those that attempt to prevent death. Regulation between these two opposing behaviours is necessary to ensure survivability—and gene regulatory networks and metabolic networks are the mechanisms that provide this regulation. We know that such regulatory behaviour is encoded in an organism's genes. The question is, how is it encoded? The understanding of this encoding can help with the development of an artificial organism, for example an autonomous robotic system; whereby the robot will have the ability to autonomously regulate the switching between the opposing behaviours using this encoded mechanism, in order to ensure its sustainable and continuous system operations. This paper aims to look into the properties of an artificial bio-chemical network consisting of a genetic regulatory network and a metabolic network that can provide these capabilities.

Keywords: artificial organisms, artificial biochemical networks, robotics.

1 Introduction

An organism is constantly faced with the dilemma of balancing two principal opposing behaviours: either to perform predatory behaviours or to avoid becoming a prey (anti-prey behaviours). Examples of predatory behaviour include working, fighting and killing in order to gather as much food as possible. Anti-prey behaviours are those that avoid potentially deadly situations or conditions, for example hibernation. If an organism only performs predatory behaviour, it may exhaust itself and have insufficient energy to continue living. If an organism only concentrates on anti-prey behaviours, the organism will be faced with the likelihood of insufficient energy and therefore be unable to survive. An organism must therefore regulate between these two behaviours to ensure survivability.

The authors of [1] describe that the biochemical networks within the body underline the functional and structural complexities within biological organisms. They state that the functionalities of biological organisms emerge from the orchestrated activities of the biochemical networks operating within individual cells.

M.A. Lones et al. (Eds.): IPCAT 2012, LNCS 7223, pp. 200–211, 2012.
© Springer-Verlag Berlin Heidelberg 2012

If we were to develop an artificial organism, for example an autonomous robotic system, (i) how can we categorise the behaviours of the robot into predatory and anti-prey behaviours and (ii) how are these behaviours encoded by its genes. Also (iii) is there a mechanism that regulates and maintains a balance between these two opposing behaviours in order to ensure sustainable and continuous system operations?

This paper attempts to answer these question by investigating how the regulatory balance is achieved for an artificial (single-celled) organism whose behaviour is governed by coupled artificial genetic and metabolic networks (described in [1] and [2]). This paper is further divided into four sections: Section 2 describes what are the predatory and anti-prey behaviours. Section 3 introduces the artificial biochemical network (ABN) and section 4 discusses how the regulation of these two behaviours is captured by the ABN. Section 5 concludes the paper.

2   Predatory and Anti-prey Behaviours

We propose the categorisation of behaviours into the two opposing predatory and anti-prey categories of behaviours. Such behaviours are observed in nature, both in multicellular and single-celled organisms.

2.1   Single Celled Organisms

By way of example, we describe the behaviours performed by two types of single-celled organisms, the flora bacteria and bacteriophage lambda:

Bacteriophage lambda survives in two phages. When it infects a host, for example *E. Coli* bacteria, it is either lysic or lysogenic. Lysogenic, the incorporation of the bacteriophage's DNA into *E. Coli*'s genome, ensures the survivability of its DNA through the evolution of the *E. Coli* bacteria. Lysic causes the destruction of its host to create more offspring. The bacteriophage becomes lysic when it senses its host is unhealthy [3]. We can thus characterise lysogenous as anti-prey phage (behaviour) and lysis as predatory phage.

Flora bacteria, in turn, ensure their survivability by continually resisting their destruction within the hostile environment of the human gut. Flora bacteria perform this functionality by creating a symbiotic relationship with the human gut. The flora bacteria must also ensure they have sufficient energy, by capturing the available energy resources within the human gut, in order to reproduce (and therefore evolve) at a high rate and prevent the eradication of their species [4]. The symbiotic behaviours can be categorised as anti-prey behaviour (since a human can survive with no flora bacteria) and to eat and reproduce at a high rate can be considered as predatory behaviours.

2.2   Early Multi-cellular Organisms

Early multi-cellular organisms or colonies of single-celled organisms coordinate their individual behaviours for the greater good of the collective. An example of early multi-cellular organism is the social amoeba *Dictyostelium discoideum* [5]. The collective and singular behaviour of the single-celled organisms can also be categorised in these two categories of behaviours. The organism(s) must find food (fight, kill and eat) in order to have sufficient (collective) energy to construct shapes or move in the environment. This allows the colony to escape and/or avoid danger from other predators or the effects of its environment. The former can be characterised as predatory behaviour and the later as anti-prey behaviour.

Sufficient energy is also necessary for reproduction. To ensure sufficient energy for the collective, unhealthy cells will scarify itself (apoptosis) for the greater good of the collective. This behaviour can also be characterised as anti-prey behaviours [5].

2.3   Regulation between the Two Opposing Behaviours

Because of the simplicity of the single-celled organisms, the regulation of the single celled organism's behaviours is directly provided by the genetic regulatory network or GRN [5], [6]. Therefore, this blueprint that governs the characterisation and the switching between these two categories of behaviours can help us with (i) the characterisation of the system's behaviours for an artificial organism into the two opposing categories of behaviours: predatory and anti-prey, and (ii) to allow the regulation that controls the switching between these two categories of behaviours to ensure stable system (equilibrium) is achieved and maintained.

The authors of [2] noted that the most significant interaction between biochemical networks in biological cells is the manner in which the genetic network controls when and where proteins are expressed, and thereby determines which enzymes are present in the metabolic network and hence which reactions can take place within a cell. In effect, the genetic network is able to reconfigure the cell's processing machinery over the course of time.

Therefore, we proposed the use of an artificial bio-chemical network to help answer our presented questions.

3   Artificial Biochemical Network

The coupled artificial biochemical network (ABN) model presented in [1] comprises an artificial genetic regulatory network (AGN) coupled to an artificial metabolic network (AMN) using a function $X : g_C \to E$, where $g_C \subseteq G$ is the set of enzyme coding genes (each enzyme (e_i) coupled to a single gene (g_i)). Coupling is achieved by giving each enzyme an expression level Ei set to the expression level of the gene it is coupled with, $\forall (g_i, e_j) \in X : Ej := Gi$. This expression level then determines the relative influence of each enzyme when calculating the new concentration of a chemical, $X : E \to C$. This captures the idea that changes in the genetic network lead to changes in the balance between competing pathways in a metabolism [1].

The genetic coding for the ABN (Fig. 1) originally presented in [1] and [2] is used. The sigmoid function $f(x) = (1 + e^{s \sum wx+b})^{-1}$ is used as the regulatory function $f_i(x)$ (that determines the gene expression level Gi) and metabolic reaction $m_i(x)$ (that determines the chemical expression C_i). wx is the weighted sum inputs to g_i and C_i, s is the slope and b is the bias ($b = 0$). The sigmoid function was chosen because this function allows easy inference of whether a gene and/or enzyme is generating predatory or anti-prey behaviours.

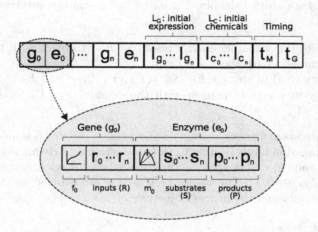

Fig. 1. Genetic encoding of an artificial bio-chemical network, taken from [1]. $s_i :=$ $Gi := Ei$ and $p_i := C_i$.

4   How Are the Behaviours Encoded within the Networks?

To investigate how behaviours are executed and regulated by the ABN, a population of 100 ABNs were evolved using a multi-objective evolutionary algorithm. Each ABN is evolved to allow the artificial (single-celled) organism the ability to perform the three basic survival behaviours:

1. When an organism is exhausted, it requires "food" to replenish its demand for energy (E_D). When in this state, the organism must make use of the available energy resources in the environment (E_R), in order to meet with the energy demand of the "body".
2. If the environment provides more "food" than it requires, the organism should not consume the excess energy resources but rather to save the excess resources (S_E) for later usage; for example at a time of limited "food" availability. This is because over-indulgence can cause the organism to become lethargic, and limits its capabilities. A failure to save the excess energy resources can cause "starvation" (at times of limited "food") and thus limit the organism's ability to survive.

3. At times of limited food, the organism should use the previously stored energy resource (E_S) to meet with its demand for energy.

For the experiments, E_D and E_R are visualised as sinusoidal values with a minimum level of 0 indicating zero energy demand and energy availability and a maximum level of 1 indicating maximum energy demand and energy availability (illustrated in Fig 4) and are the inputs signals to the ABN.

In summary, the ABN for the artificial organism (the robot) will carry out the following regulatory behaviours in order to ensure the survivability of the organism:

1. make use of the available energy resource (E_R) to meet with the required energy demand (E_D), $U_E = E_R.C_{O2}$,
2. save energy (S_E) in storage E_S, $S_E = E_R.(1 - C_{O2})$ and $E_S = E_S + S_E$.
3. use energy in E_S to help meet with the demand E_D, $U_E = E_S.C_{O1}$ and $E_S = E_S - U_E$.

Each ABN consists of 6 genes, comprising 54 weights (w) and 6 slopes (s), coding for 6 gene expression levels leading to the production of 2 chemical outputs C_{O1} and C_{O2} ($6w$ and $2s$) that govern the three described behaviours. $w \in [-1, 1], s \in [-1, 1]$ and $C_{Oy} \in [0, 1]$. The values of w and s are evolved so that the best evolved ABNs achieved the fitness objectives of:

1. $E_D \approx (U_E + U_S)$.
2. $S_E > 0$.

4.1 The Evolved ABN

After investigating the properties of the evolved ABNs in the population, the following observations were made:

1. to make use of the available energy in E_S to meet with the required E_D is considered as anti-prey behaviour. This is because the best evolved ABNs (ABN B and ABN E - Fig. 8) have $C_{O1}(x) < 0.5$. [1].
2. to use E_R to meet with the required E_D is considered as as predatory behaviour because $C_{O2}(x) > 0.5$.

These observations show that behaviours of the artificial organism can be placed into the two categories: (i) predatory and (ii) anti-prey; and that the genes' and enzymes' w and s values encode these behaviours. If incorrect values of w and s are used within the AGN, the artificial organism has a lesser inability to meet with E_D.

If we compare ABN A and ABN B (see Fig 4), ABN B is more effective at meeting with E_D, since $(U_E + U_S) - E_D$ is larger for ABN A than ABN B. This is because, if we follow the previously stated rules:

---

[1] $C_{On}(x) = m_n(x) = (1 + e^{(s \sum w(x))})^{-1}$.

1. If the gene is always producing $g_y(x) > 0.5$, this gene is considered a predatory gene.
2. If the gene is always producing $g_y(x) < 0.5$, this gene is considered a anti-prey gene.

ABN B has 3 predatory genes, 2 anti-prey genes and 1 neutral gene[2], in comparison to ABN A that has 4 predatory genes, 0 anti-prey genes and 2 neutral genes. The additional anti-prey genes for ABN B allows the organism to perform the anti-prey behaviour of using available energy in E_S to meet the required E_D more effectively. Figure 2 illustrates that solutions with similar network characteristics have greater ability to meet with E_D.

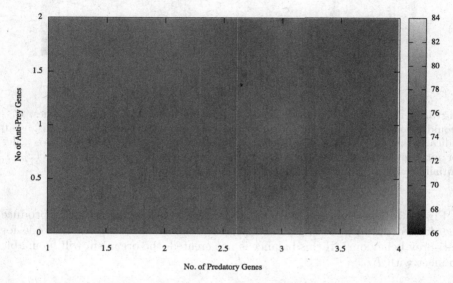

Fig. 2. The relationship between the number of anti-prey and predatory genes within the AGN and the organism's fitness value. Each AGN consists of 6 genes. Three types of gene characteristics were found. If a gene is constantly producing $g_y(x) > 0.5$, this gene is considered a predatory gene. If a gene is constantly producing $g_y(x) < 0.5$, this gene is considered a anti-prey gene. If the gene oscillates with a mean of 0.5, this is considered a neutral gene. The lower the fitness value the better the AGN is at meeting its E_D. AGN B and E have fitness values of 64. The figure shows that the organism requires all the predatory genes, anti-prey genes and the neutral genes in order to meet its objectives, with the best network configuration of 3 predatory genes, 2 anti-prey genes and 1 neutral gene within the AGN. Note: the 1/2 values in the x- and y-axis should be ignored.

Figure 2 illustrates that there are two best network configurations:

1. 3 predatory genes, 2 anti-prey genes and 1 neutral gene, or
2. 3 predatory genes, 1 anti-prey gene and 2 neutral genes

---

[2] its values oscillate with a mean of 0.5.

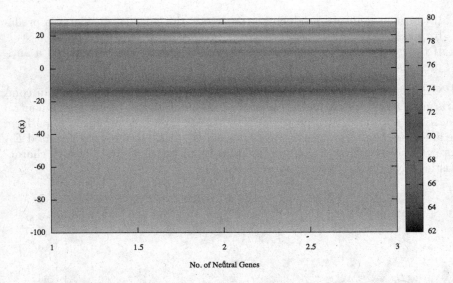

Fig. 3. If there are 3 predatory genes, 1 anti-prey gene and 2 neutral genes, C_{O2} should encode for anti-prey behaviour to allow the organism to meet with E_D. $c(x) > 0$ indicates that C_{O2} codes for predatory behaviour and $c(x) < 0$ indicates that C_{O2} codes for anti-prey behaviour. If $C_{O2} > 0.5$, $c(x) = c(x) + 1$ and if $C_{O2} < 0.5$, $c(x) = c(x) - 1$. Initially $c(x) = 0$. Note: the 1/2 values in the x-axis should be ignored.

Figure 3 shows that in order for the second network configuration to produce the desired output, its C_{O2} should produce more < 0.5 values (and code for anti-prey behaviour). If this balance is not created, the organism will be unable to meet with E_D.

4.2   Assigning the w and b Values

Comparing ABN G with ABN F, Fig. 6 shows that the more similar $+w$ values (and the higher their magnitude) are for a gene, the higher the likelihood the gene will be turned on ($g_1 - g_3$ of AGN G). The more similar are the $-w$ values (and the higher their magnitude) for a gene, the higher the likelihood that the gene will be turned off; for example C_{O1} and C_{O2} of AGN G (Fig. 5). Also, the less varying the w values (for example the large number of w with $+0.2 < w < 0.2$ for g_4 of ABN F), the less likely that the gene will become an oscillator. Similar observations were made when comparing AGN E and AGN H (Fig. 7).

These results agree with the observations presented in [3] and [6], where the authors state that the cell behaviour is produced through the interplay between positive regulation and negative regulation, creating a behavioural toggle switch with sustained oscillation [3], [6].

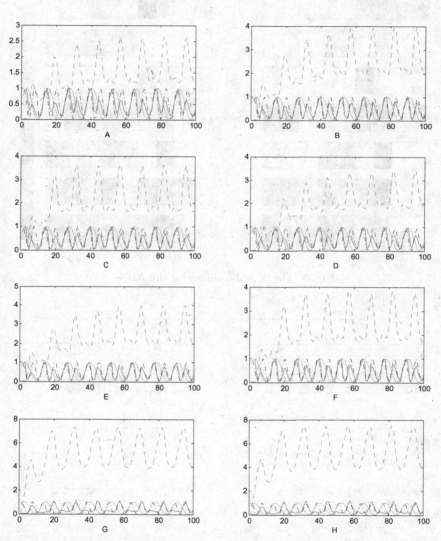

Fig. 4. The outputs produced by eight evolved ABNs. Red line $= U_E + U_S$, blue line $= E_D$, green line $= E_R$ and the magenta line $= S_E$.

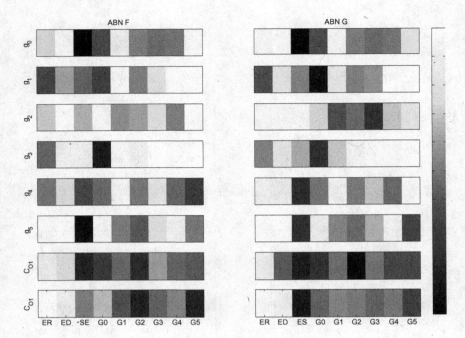

Fig. 5. The w and b values for the ABNs

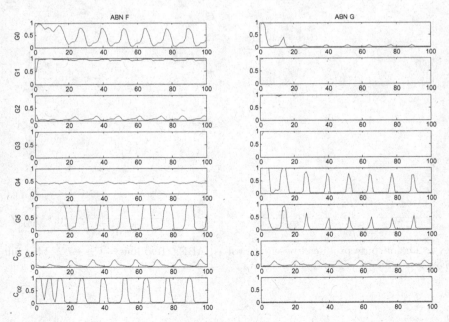

Fig. 6. The genes (g_x) and chemical expressions (C_{Oy}) for the ABN. C_{O2} indicates how E_R is to be used and C_{O1} coordinates how S_E is to used.

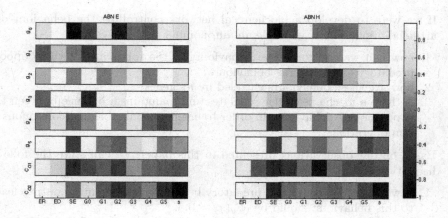

Fig. 7. The w and b values for the ABNs

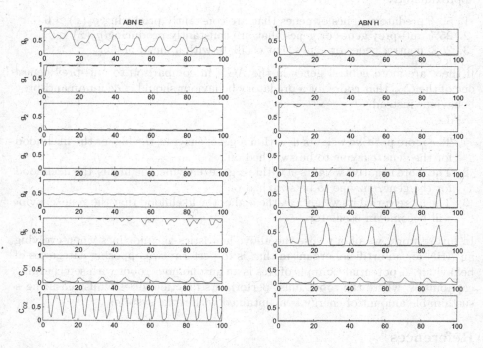

Fig. 8. The genes (g_x) and chemical expressions (C_{Oy}) for the ABN. C_{O2} indicates how E_R is to be used and C_{O1} coordinates how S_E is to be used.

5   Conclusion

If we were to develop a biochemical network controlling the behaviour of an artificial organism, for example an autonomous robotic system:

1. how can we categorise the behaviours of the robot into the two opposing predatory and anti-prey behaviours?
2. how are these behaviours encoded by its genes?
3. is there a mechanism that regulates and maintains a balance between these two opposing behaviours in order to ensure sustainable and continuous system operations?

Based on our observations presented in this paper, we can state the following heuristics:

1. If we wish to encode for a predatory behaviour, the chemical output leading to this behaviour should be: $C_y(x) > 0.5$.
2. Similarly, if we wish to encode for an anti-prey behaviour, the chemical output should be: $C_y(x) < 0.5$.

Furthermore, to allow the genes to act as a toggle switch which enables oscillation between the two opposing categories of behaviour, the AGN should consist of approximately:

1. 50% predatory genes or genes that are constantly producing $g_y(x) > 0.5$.
2. 25% anti-prey genes or genes that are constantly producing $g_y(x) < 0.5$.
3. 25% neutral genes or genes that oscillate with a mean of 0.5.

If there are more neutral genes in the AGN in comparison to anti-prey genes, one of the C_{Oy} that codes for a predatory behaviour should code for an anti-prey behaviour instead.
    Also,

1. The more positive weights w within a gene/enzyme, the higher the likelihood for the gene/enzyme to be switched off.
2. The more negative weights w within a gene/enzyme, the higher the likelihood for the gene/enzyme to be switched on.
3. The more varied its weights w, the higher the likelihood that the gene/enzyme will become an oscillator.

In future work, we plan to use the above heuristics as guidelines when evolving an ABN for an artificial organism that is to perform two opposing categories of behaviour. A potential example of this is an autonomous robot with self charging capabilities, where the robot must perform its functionalities whilst ensuring a sustainable amount of energy is maintained within the system.

References

1. Lones, M.A., Tyrrell, A.M., Stepney, S., Caves, L.: Controlling Legged Robots with Coupled Artificial Biochemical Networks. In: Proc. European Conference on Artificial Life, ECAL (August 2011)

2. Lones, M.A., Tyrrell, A.M., Stepney, S., Caves, L.: Controlling Complex Dynamics with Artificial Biochemical Networks. In: Esparcia-Alcázar, A.I., Ekárt, A., Silva, S., Dignum, S., Uyar, A.Ş. (eds.) EuroGP 2010. LNCS, vol. 6021, pp. 159–170. Springer, Heidelberg (2010)
3. Hasty, J., Isaacs, F., Dolnik, M., McMillen, D., Collins, J.J.: Designer gene networks: Towards fundamental cellular control. Choas 11(1), 207–220 (2001)
4. Ley, R.E., Peterson, D.A., Gordon, J.I.: Ecological and evolutionary forces shaping microbial diversity in the human intestine. Cell 124, 837848 (2006)
5. Tsuchiya, H.M., Drake, J.F., Jost, J.L., Fredrickson, A.G.: Predator-Prey Interactions of Dictyostelium discoideum and Escherichia coli in Continuous Culture. Journal of Bacteriology 110(3), 1147–1153 (1979)
5. de Jong, H.: Modeling and simulation of genetic regulatory systems: a literature review. Journal of Computational Biology 9, 67103 (2002)
6. Wolf, D.M., Eeckman, F.H.: On the relationship between genomic regulatory element organization and gene regulatory dynamics. Journal Theoretical Biology 198, 167–186 (1998)
7. Balaji, S., Babu, M.M., Aravind, L.: Interplay between network structures, regulatory modes and sensing mechanisms of transcription factors in the transcriptional regulatory network of E. Coli. Journal of Molecular Biology 372(4), 1108–1122 (2007)
8. Tyson, J.T., Chen, K., Novak, B.: Network Dynamics and Cell Physiology. Nature 2, 908–916 (2001)

Closing the Gap between Life and Physics

Ron Cottam, Willy Ranson, and Roger Vounckx

The Living Systems Project, Department of Electronics and Informatics,
Vrije Universiteit Brussel, Pleinlaan 2, 1050 Brussels, Belgium
life@etro.vub.ac.be

Abstract. Examination of the scalar properties of living organisms and the electronic configuration of crystalline structures suggests that similar modeling may be used for both. This paper comments on individual and common properties of the two systems and draws a comparison between them. Both exhibit multiple scales and a global 'overview' of their scalar properties. We conclude that the analogy may provide a fruitful route towards modeling living organisms.

Keywords: Life, scale, hyperscale, solid state physics.

1   Introduction

The results of many and varied investigations over the last century have only gone to show that there is a wide gulf between our understandings of living systems and conventional physics. A major distinguishing feature is that while physical materials exhibit a degree of informational difference across their scales, it is neither so complex nor functionally integrated as in living systems. A defect in the investigator's toolbox has been the lack of a self-consistent study of scale in natural hierarchical structures. This, however, is now available [1, 2]. Surprisingly, even simple crystalline materials show informational differences across scales. A case in point is the collection of zinc-blende structured crystals which comprise the chemical group IV, the III-V and the II-VI materials [3]. However, this difference is only less than 1%, or a few %, while organisms exhibit radically different informational properties between adjacent scales, for example between tissues and their constituent cells.

The basic difficulty in moving from one informational scale to a next higher one is that while information may be gained in the operation it is at the expense of loss of information associated with the initial scale. An apposite analogy is the difficulty of actually carrying out the operation 1+1=2, which is itself hierarchical in a similar manner: degrees of freedom are lost in the operation which yields the higher order 2. This means that although it may be possible to upscale from 1+1 to 2 (which is, in reality, a prior definition by rule), it is impossible to correctly return downscale (e.g. for 1+2=3, it is impossible from 3 to know if the lower scale constitutes 1+2, 2+1 or even 1+1+1). This is the same problem which makes memory necessary in a Boolean computer, where the gates effectively 'throw away' information at each stage. In passing, it is instructive to note that the primary function of the clock in a Boolean computer is to eliminate any local-to-and-from-global communication, making the

M.A. Lones et al. (Eds.): IPCAT 2012, LNCS 7223, pp. 212–215, 2012.
© Springer-Verlag Berlin Heidelberg 2012

establishment in a Boolean computer of any phenomenon with global properties impossible – e.g. intelligence; consciousness.

2   Linking Life and Physics

So, scale is a tricky beast to deal with - particularly in living systems. This is the major source of our lack of comprehension linking life and physics. In this paper we attempt to begin constructing a bridge between the two, by reference to the currently most detailed physical modeling scheme: that describing the solid state physics of the integrated circuits which make up conventional computers. An initial objection which comes to mind is that there is almost-perfect long range atomic order in the Si or GaAs which are usually used in integrated circuits.

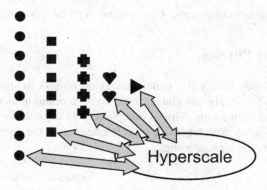

Fig. 1. Integration of the scales of a unified natural hierarchy into its hyperscalar identity

Surely the lack of similar order in biological materials makes such a comparison fatuous? Well, not necessarily. First, many biological molecules exhibit a degree of similar order – for example the lipid pdmpg. Second, it may be that the approximate spatial repetition of cells in tissue is just enough to provide long-range coupling. Third, and most importantly, it is not an exact model we are looking for here, it is a general idea which may stimulate new representations of the adhesive coupling between scales in living systems.

To cut a (very) long story short, inter-scalar coupling at a local level in biomaterials is dependent on global properties [4]. Amongst other reasonings, this follows from the observation that no matter how disparate the scales of an organism may be, the organism functions self-consistently at a global level – the collection of individual scales is integrated into a single 'self-representation', referred to as hyperscale, which constitutes its global nature [5] (see Fig. 1). This provides us with a guide in searching for related models in more conventional physical domains, and it is notable that solid state physics not only exhibits a series of energetically isolated 'scales', but that their associated Brillouin zones may be folded into a single description which resembles hyperscale.

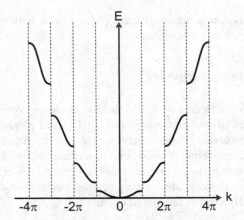

Fig. 2. 'Real' solutions to the Kronig-Penney equation for electron energies in a crystal

3   Solid State Physics

Free electrons in solids follow the same rules as do objects in space: their energies and momenta may be simply calculated, resulting in a continuous quadratic relation between energy and momentum. This is complicated in crystals by the repetitive appearance of atomic ions, which provides a reflection when the energetic wavelength of electrons matches the ionic spacing.

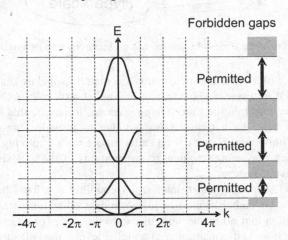

Fig. 3. The Brillouin zones folded back into the central $-\pi$ to $+\pi$ region

Following the archetypical model derived by Kronig and Penney [6], the net result is to split the free-electron energy-momentum parabola into a set of 'permitted' regions separated by 'forbidden' gaps – each permitted region constituting a Brillouin zone, as shown in Fig. 2. These regions are related in terms of a sin/cos equation, and the entire parabola may be folded in on itself to occupy solely the central $-\pi$ to $+\pi$

region of the momentum axis (more precisely, between $-\pi$ and $+\pi$ in 'k-space', where k is the electronic wave number – see Fig. 3). Consequently, free-electron properties in crystals provide an analogy to the multi-scalar/hyperscalar properties of biological multi-scalar systems and of life [7].

4   Conclusion

It would clearly be going too far to attribute this exact mechanism to biological inter-scalar relationships. The likelihood of a similarly purely quantum mechanical solution is very far-fetched. However, it has recently come to light that quantum logic may operate/be operated in the absence of quantum systems *per se* [8], and Karl Pribram has mooted the possibility of quasi-quantum wave collapse in the axonite mesh of neural systems [9]. It may be that our understanding of the relationships between Newtonian and quantal mechanics is far too limited, and that biological materials may indeed be characterized by quantum-like processes, if not by quantum mechanical processes themselves.

References

1. Cottam, R., Ranson, W., Vounckx, R.: Autocreative hierarchy I: structure - ecosystemic dependence and autonomy. SEED Journal 4, 24–41 (2004)
2. Cottam, R., Ranson, W., Vounckx, R.: Autocreative Hierarchy II: dynamics - self-organization, emergence and level-changing. In: Hexmoor, H. (ed.) International Conference on Integration of Knowledge Intensive Multi-Agent Systems, pp. 766–773. IEEE, Piscataway (2003)
3. Cottam, R., Saunders, G.: The Elastic Constants of GaAs from 2K to 320K. Journal of Physics C: Solid State Physics 6, 2015–2118 (1973)
4. Cottam, R., Ranson, W., Vounckx, R.: A diffuse biosemiotic model for cell-to-tissue computational closure. BioSystems 55, 159–171 (2000)
5. Cottam, R., Ranson, W., Vounckx, R.: Living in hyperscale: internalization as a search for unification. In: Wilby, J., Allen, J.K., Loureiro-Koechlin, C. (eds.) Proceedings of the 50th Annual Meeting of the International Society for the Systems Sciences, paper #2006-362, ISSS, Asilomar, CA, pp. 1–22 (2006)
6. de L. Kronig, R., Penney, W.G.: Quantum Mechanics of Electrons in Crystal Lattices. Proceedings of the Royal Society (London) A130, 499 (1930)
7. Cottam, R., Ranson, W., Vounckx, R.: Towards Cross-modeling between Life and Solid State Physics. In: Integral Biomathics: Tracing the Road to Reality. Springer, Heidelberg (2011) (in publication)
8. Schroeder, M.: Quantum Coherence without Quantum Mechanics in Modeling the Unity of Consciousness. In: Bruza, P., Sofge, D., Lawless, W., van Rijsbergen, K., Klusch, M. (eds.) QI 2009. LNCS (LNAI), vol. 5494, pp. 97–112. Springer, Heidelberg (2009)
9. Pribram, K.: Proposal for a quantum physical basis for selective learning. Presented at the 4th International Conference on Emergence, Complexity, Hierarchy and Order, Odense, Denmark, July 31-August 4 (2001)

Algebraic Analysis of the Computation in the Belousov-Zhabotinksy Reaction

Paolo Dini[1,2], Chrystopher L. Nehaniv[1],
Attila Egri-Nagy[1,3], and Maria J. Schilstra[1]

[1] Royal Society Wolfson BioComputation Research Lab
Centre for Computer Science and Informatics Research
University of Hertfordshire, United Kingdom
{p.dini,c.l.nehaniv,m.j.1.schilstra}@herts.ac.uk
[2] Department of Media and Communications
London School of Economics and Political Science
London, United Kingdom
[3] School of Computing and Mathematics
University of Western Sydney, Parramatta, New South Wales, Australia
a.egri-nagy@uws.edu.au

Abstract. We analyse two very simple Petri nets inspired by the Oregonator model of the Belousov-Zhabotinsky reaction using our stochastic Petri net simulator. We then perform the Krohn-Rhodes holonomy decomposition of the automata derived from the Petri nets. The simplest case shows that the automaton can be expressed as a cascade of permutation-reset cyclic groups, with only 2 out of the 12 levels having only trivial permutations. The second case leads to a 35-level decomposition with 5 different simple non-abelian groups (SNAGs), the largest of which is A_9. Although the precise computational significance of these algebraic structures is not clear, the results suggest a correspondence between simple oscillations and cyclic groups, and the presence of SNAGs indicates that even extremely simple chemical systems may contain functionally complete algebras.

1   Introduction

In self-organising systems, the "self" or autonomous aspect is provided by the fall towards equilibrium, which serves as the driver or energy source. As a consequence, a system that needs to maintain self-organising behaviour indefinitely must be open since, if it were closed, once it had reached equilibrium it would stop functioning. Therefore, in order to keep going it must be open and connected to a source of (free) energy that can keep it 'far from equilibrium', to use Prigogine's famous phrase [19], even whilst it is continually falling towards it. The Belousov-Zhabotinsky (BZ) reaction has been studied extensively [20] because it was the first reaction to exhibit sustained oscillations even in an isolated system, although they do die down eventually. Before Belousov's discovery in the 1930s and Zhabotinsky's confirmation of the phenomenon in the 1960s,

M.A. Lones et al. (Eds.): IPCAT 2012, LNCS 7223, pp. 216–224, 2012.

species concentrations were believed to vary monotonically unless driven by a periodic forcing function. In a constant-flow reactor the oscillations are periodic and can be sustained indefinitely, as long as the inflow and outflow are kept constant. This qualifies the BZ reaction as a system far from equilibrium. In this paper we analyse a simplified ordinary differential equation (ODE) model of the BZ reaction, the almost equally famous "Oregonator" model, developed by Field and Noyes at the University of Oregon [11].

We compare the structure and behaviour of a very simple system inspired by the Oregonator model of the BZ reaction from the different viewpoints of systems biology and algebraic automata theory. In particular, we focus on its oscillatory behaviour. Although the computational properties of chemical oscillations are not clear, the fact that we are very familiar with them at both an intuitive and a mathematical level makes them a useful reference system when attempting to decipher the computational significance of the algebraic structures uncovered in the corresponding finite state automata, as we discuss below. This was, in fact, the main motivation for selecting the BZ reaction as an object of study. Thus, this work aims to merge two research traditions: dynamical systems theory rooted in physics and informing much of modern-day systems biology, and theoretical computer science rooted in algebraic automata theory [16].

2   Discretisation

To be able to analyse the (Oregonator model of the) BZ reaction from these two different perspectives we must find a way to discretise it. A good way to achieve this is with a Petri net (PN), although Kauffman's Boolean networks [15,7] and Rhodes's reaction digraphs [22] are also useful possibilities, all amenable to algebraic automata-theoretic methods. The PN notation, invented to describe interaction and transformation in discrete distributed systems [21,6], is highly suitable to depict the structure of biochemical reaction networks at the level of interaction between molecules. In combination with kinetic information, PN models are useful tools in the derivation of the coupled ODE systems that describe the dynamic behaviour of these networks [23]. Once such a PN has been obtained, it is straightforward to derive a finite-state automaton by treating each possible marking of the PN as a different state of the automaton. In order to obtain a finite state automaton the number of tokens is bounded, and the bound is called the "capacity" of the place. This can be viewed as discretising concentration levels or bounding the number of molecules of each type. Since the resulting number of states can be very large, normally the markings-to-states mapping is done for a specific choice of initial conditions, which yields a subset of the global automaton of all possible states.

3   Krohn-Rhodes Theory

The Krohn-Rhodes prime decomposition theorem for finite automata [16] has been discussed, explained, and applied in a large number of books and articles since the theorem was published in 1965 ([1] and many others). In 1967 Zeiger [26] proved a variant, called holonomy decomposition (HD), according to which any finite automaton can be decomposed into a cascade of permutation-reset automata arising from a study of how inputs act on certain subsets of the powerset of the state set. In other words, the state transitions of the component automata can only be either permutations (of certain subsets of subsets) of the state set or resets (Cases 1a and 1b in Figure 1, respectively). The permutation-reset automata can then be further decomposed into (finite and discrete) groups and two-state resets (also known as flip-flops). Finally, using the Jordan-Hölder theorem each group can be further subdivided into a sequence of simple groups, known as its composition factors, recovering the Krohn-Rhodes decomposition into irreducible atomic groups (simple groups) and combinatorial semigroups (cascades of banks of flip-flops). The HD has continuously been improved in efficiency over the years (e.g., [10,13,3,18,4], finally leading to computer algebraic realisation ([9], which has more recently been reimplemented in GAP [25] as SgpDec [8]), making possible the decomposition and analysis of structures previously well beyond human capacity to analyse.

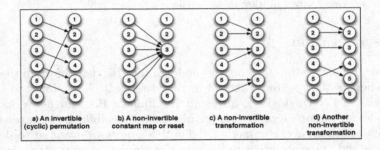

a) An invertible (cyclic) permutation   b) A non-invertible constant map or reset   c) A non-invertible transformation   d) Another non-invertible transformation

Fig. 1. Different kinds of transformations of 6 states (inspired by [18])

HD diagrams, used to visualise the decomposition, do not show how the groups act explicitly, but indicate the presence of different groups at the various levels of the cascade. These groups indicate the presence of 'local pools of reversibility' [22], but it is still unclear what *algorithmic* significance the transitions these groups induce might have. Further, they show up in the decomposition of automata derived from metabolic and regulatory pathways. Because every group is associated with one or more symmetries and because biological systems exhibit – in fact, depend on – many symmetries in their structure and behaviour, it seems inescapable that the groups embedded in the HD of biological automata have something to do with their self-organising properties. We now look at the BZ system in more detail.

4   Analysis of BZ Reaction

The BZ reaction has been analysed by many people over the last 50 years. In addition to the Field, Noyes, and Körös works already cited a useful reference is Scott [24], on whom we mostly rely. The size of the HD tends to increase exponentially with automaton size. Thus, although SgpDec has made it possible to analyse systems that are immensely greater than what could previously be done only by hand, we still need to keep the systems analysed as small as possible. Therefore, although the Oregonator model is already a significant simplification relative to the full set of chemical equations of the BZ system, we had to simplify it further in order to bring the corresponding automaton to a size amenable for algebraic analysis. We now provide a brief summary of the original model and corresponding ODE system in order to explain and justify the simplifications effected. The Oregonator chemical equations are the following [24]:

$$A + Y \rightarrow X + P \qquad\qquad rate = k_3 AY \qquad\qquad (1)$$
$$X + Y \rightarrow 2P \qquad\qquad rate = k_2 XY \qquad\qquad (2)$$
$$A + X \rightarrow 2X + 2Z \qquad\qquad rate = k_5 AX \qquad\qquad (3)$$
$$2X \rightarrow A + P \qquad\qquad rate = k_4 X^2 \qquad\qquad (4)$$
$$B + Z \rightarrow (1/2)fY \qquad\qquad rate = k_c BZ \qquad\qquad (5)$$

X, Y, and Z correspond to the three compounds that undergo periodic oscillations under steady-state boundary conditions, meaning inside the continuous-flow stirred-tank (CFST) reactor. A, B, and P, by contrast, do not vary as a function of time, a consequence of the CFST reactor setup. The three colours Red, Green and Blue shown correspond to the three processes the BZ reaction is conceptually divided into.[1] Table 1 describes the variables in question.

Table 1. Summary of Oregonator variables

Oregonator Variable	Chemical Compound	Chemical Symbol	Association with BZ Reaction Process
X	Bromous Acid	$HBrO_2$	Process B (reduces X, generates Z)
Y	Bromide Ion	Br^-	Process A (reduces Y, generates X)
Z	Cerum 4	$Ce(IV)$	Process C (reduces Z, generates Y)
A	Bromate	BrO_3^-	All 3 processes
B	Malonic Acid	$CH_2(COOH)_2$	Process C
P	Hypobromous Acid	$HOBr$	Process A and C

---

[1] Using the potentially confusing accepted notation, the names of the processes have nothing to do with the letters A, B, and P assigned to three of the compounds.

From Eqs. 1-5 the rate equations are easily derived as a set of three ODEs:

$$\frac{dX}{dt} = k_3 AY - k_2 XY + k_5 AX - 2k_4 X^2 \tag{6}$$

$$\frac{dY}{dt} = -k_3 AY - k_2 XY + 1/2 f k_c BZ \tag{7}$$

$$\frac{dZ}{dt} = 2k_5 AX - k_c BZ, \tag{8}$$

where the colours indicate the contribution to the rates of change of the three variables due to the three processes A, B, and C. Since our objective is to reach an intuitive understanding of the computational significance of the algebraic structure of automata derived from biochemical systems, we now proceed to make a number of radical simplifications to this system, a step which we believe to be necessary at this early stage in the analysis. In fact, since, as we shall see below, the HD results can be very difficult to interpret even with very simple systems, it is important to start with the simplest possible system.

In order to see behaviour that is close to a non-equilibrium dynamical system's we needed each place to have a capacity of at least 4 tokens, so we reduced the number of places to the smallest number possible, i.e. 3. The justification for eliminating the three compounds A, B, and P lies in the fact that for a CFST reactor they remain constant. Thus, including them greatly increases the state space of the automaton derived from the PN in spite of their not contributing to the dynamics we are investigating. Furthermore, we eliminated also k_2 and k_4 in Eqs. 6-8. The motivation is that although they are important for reproducing the shape of the BZ oscillations, the oscillations themselves can be generated with a simpler system, which is preferable for now. Similarly, the factor of 2 in front of the first term on the right-hand side of Eq. 8, which comes from the factor of 2 in front of Z in Eq. 3, is also ignored in constructing the PN. This will need to be brought back in at a later stage because it is responsible for the characteristically fast growth of the Z compound.

The problem with these modifications is that the resulting system is so different from the original BZ system that it may not even oscillate. This is remedied by artificially introducing inhibition, which is applied cyclically around the three active compounds. Unfortunately once this step is taken it becomes impossible to compare directly the resulting PN to the original ODE system, even if simplified. However, since the average place concentrations obtained with a stochastic PN simulator will converge to the ODE results as the number of runs approaches infinity [12], we can still analyse the resulting system as if we *did* know the governing equations. The result of all these simplifications and modifications is an extremely simple and highly symmetrical PN, shown in Figure 2. A weight n of the inhibition arc between, for example, Place Y and Transition t_1 means that t_1 is inhibited if Y contains n or more tokens.

Figure 2 shows the GAP input file prepared for our PN package [5] and the 15-state automaton corresponding to the initial condition $(0, 4, 0)$ (States 5, 8 and 10 are not reachable and are not shown). At the bottom of Figure 2 the output of a stochastic PN simulator coded in *Mathematica* can be seen as a time series of the token values in the three places of this PN. The traces shown are the average of 500 runs. Damped oscillations are clearly visible, as well as the fact that this PN conserves mass ($1.3 \cdot 3 \approx 4$). The states of the automaton shown are the possible markings of the PN from the given initial condition. The rate constants for this example are all 1 ($K = (1, 1, 1)$). Finally, SgpDec was used to generate the HD, revealing 12 cascaded permutation-reset levels, whereby '1' indicates either an irreversible component or a trivial group, and the other groups are shown in standard notation. It is not very surprising that in this extremely simple and symmetrical example most of the levels of the decomposition are groups, and they are particularly simple (in the colloquial sense of the term) groups, since they are all cyclic groups.

The abstract concept of algebraic structure is useful for understanding mathematical theorems, but by itself it is not readily applicable. Fortunately, Krohn-Rhodes theory is intimately related to a cognitive tool with which we are intimately familiar in our daily lives: coordinatisation. Namely, the different levels of the Krohn-Rhodes decomposition of an automaton are analogous to the different positions in our positional number systems. The decomposition then becomes an expansion of a given automaton into an "abstract number system" that is defined by the automaton itself: each state is expressed as a different multi-digit "number", where each "digit" corresponds to a level in the decomposition. The significance of this insight, due to Rhodes, is that the coordinatisation perspective gives us at once powerful cognitive and calculational tools for manipulating an automaton in our mind or with possible software support, and also gives us the starting point of a general computer science methodology.

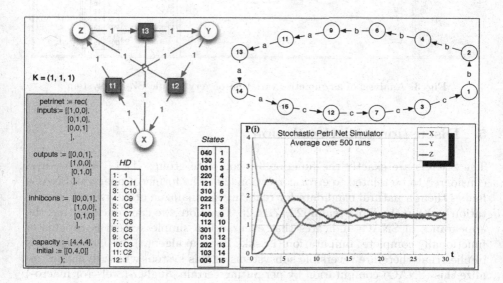

Fig. 2. Stochastic simulation and HD analysis of very simple BZ-like system

Figure 3 shows a rather different story. Here the PN is not symmetrical since one of the inhibitions is missing, and the other two have different values. The stochastic simulation shows strongly damped oscillations, and the automaton (with the same states) is more complex. The most interesting output, however, is by far the HD. The decomposition shows 35 levels, with a fair number of groups. This case is remarkable because of the presence of very large groups (S_9 has $9! = 362,880$ elements acting on subsets of the states shown, at level 14). The groups shown in red all contain SNAGs, i.e. A_9, A_8, A_7, A_6, A_5.

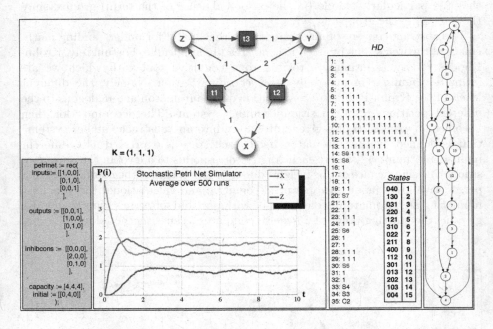

Fig. 3. Analysis of asymmetric variation of very simple BZ-like system

5   Discussion and Conclusion

The SNAGs are exactly the functionally complete groups [17,14] and are also considered to be related to error-correction [22]. The former property of SNAGs makes them a natural candidate for realizing an analogue of "universal computation" within the finite realm [22,17,14,2]. Therefore, we can conclude that the appearance of SNAGs indicates that even such a simple system is capable of functionally complete computation (i.e. like Boolean algebra). In particular, this implies that there are reversible subsystems of this system whose dynamics realize this (SNAG) computation by permuting certain "higher-level" (or macro-) states. Natural emerging research questions now are: (1) How, in detail, do oscillatory systems such as the simple BZ-like one analysed here realise this kind of (functionally complete) computation in terms of the dynamics of specific reversible subsystems? And (2) How might the finitary universal computational potential of these BZ subsystems be harnessed?

Acknowledgment. Paolo Dini gratefully acknowledges the partial support of this work by the EU project EINS, Contract number FP7-ICT-288021.

References

1. Arbib, M.A. (ed.): Algebraic Theory of Machines, Languages, and Semigroups. Academic Press (1968)
2. Dini, P., Schreckling, D.: A Research Framework for Interaction Computing. In: Colugnati, F.A.B., Lopes, L.C.R., Barretto, S.F.A. (eds.) OPAALS 2010. LNICST, vol. 67, pp. 224–244. Springer, Heidelberg (2010)
3. Dömösi, P., Nehaniv, C.L.: Algebraic Theory of Automata Networks. SIAM, Philadelphia (2005)
4. Egri-Nagy, A., Mitchell, J.D., Nehaniv, C.L.: Algorithms for the Efficient Calculation of the Holonomy Decomposition. In: Dömösi, P., Iván, S. (eds.) Automata and Formal Languages: Proc. 13th Intern. Conference AFL 2011, August 17-22. Institute of Mathematics and Informatics, College of Nyíregyháza, Debrecen, Hungary (2011)
5. Egri-Nagy, A., Nehaniv, C.L.: PN2A: Petri Net Analysis GAP Package, http://sourceforge.net/projects/pn2a/
6. Egri-Nagy, A., Nehaniv, C.L.: Algebraic properties of automata associated to Petri nets and applications to computation in biological systems. BioSystems 94(1-2), 135–144 (2008)
7. Egri-Nagy, A., Nehaniv, C.L.: Hierarchical coordinate systems for understanding complexity and its evolution with applications to genetic regulatory networks. Artificial Life 14(3), 299–312 (2008) (Special Issue on the Evolution of Complexity)
8. Egri-Nagy, A., Nehaniv, C.L.: SgpDec – software package for hierarchical coordinatization of groups and semigroups, implemented in the GAP computer algebra system, Version 0.5.19 (2010), http://sgpdec.sf.net
9. Egri-Nagy, A., Nehaniv, C.L., Rhodes, J.L., Schilstra, M.J.: Automatic analysis of computation in biochemical reactions. BioSystems 94(1-2), 126–134 (2008)
10. Eilenberg, S.: Automata, Languages, and Machines, vol. B. Academic Press, New York (1976)
11. Field, R.J., Noyes, R.M.: Oscillations in chemical systems IV. Limit cycle behavior in a model of a real chemical reaction. Journal of Chemical Physics 60(5), 1877–1884 (1974)
12. Gillespie, D.T.: A general method for numerically simulating the stochastic time evolution of coupled chemical reactions. Journal of Computational Physics 22, 403–434 (1976)
13. Holcombe, W.: Algebraic Automata Theory. Cambridge University Press (1982)
14. Horváth, G.: Functions and Polynomials over Finite Groups from the Computational Perspective. The University of Hertfordshire, PhD Dissertation (2008)
15. Kauffman, S.: The Origins of Order: Self-Organisation and Selection in Evolution. Oxford University Press, Oxford (1993)

16. Krohn, K., Rhodes, J.: Algebraic Theory of Machines. I. Prime Decomposition Theorem for Finite Semigroups and Machines. Transactions of the American Mathematical Society 116, 450–464 (1965)
17. Krohn, K., Maurer, W.D., Rhodes, J.: Realizing complex boolean functions with simple groups. Information and Control 9(2), 190–195 (1966)
18. Maler, O.: On the Krohn-Rhodes Cascaded Decomposition Theorem, http://www-verimag.imag.fr/~maler/
19. Nicolis, G., Prigogine, I.: Self-Organization in Nonequilibrium Systems. Wiley, New York (1977)
20. Noyes, R.M., Field, R.J., Körös, E.: Oscillations in Chemical Systems I. Detailed Mechanism in a System Showing Temporal Oscillations. Journal of the American Chemical Society 94(4), 1394–1395 (1972)
21. Petri, C.A.: Kommunikation mit Automaten. Schriften des IIM 2 (1962)
22. Rhodes, J.: Applications of Automata Theory and Algebra via the Mathematical Theory of Complexity to Biology, Physics, Psychology, Philosophy, and Games. World Scientific Press (2010)
23. Schilstra, M.J., Martin, S.R.: Simple stochastic simulation. In: Michael, L., Ludwig, B. (eds.) Methods in Enzymology, pp. 381–409. Academic Press, Elsevier (2009)
24. Scott, S.: Oscillations, Waves, and Chaos in Chemical Kinetics. Oxford University Press, Oxford (1994)
25. The GAP Group: GAP – Groups, Algorithms, and Programming, Version 4.4 (2006), http://www.gap-system.org
26. Zeiger, H.P.: Cascade synthesis of finite-state machines. Information and Control 10(4), 419–433 (1967)

Improving Transcription Factor Binding Site Predictions by Using Randomised Negative Examples

Faisal Rezwan[1], Yi Sun[1], Neil Davey[1], Rod Adams[1],
Alistair G. Rust[2], and Mark Robinson[3]

[1] School of Computer Science, University of Hertfordshire, College Lane, Hatfield,
Hertfordshire AL10 9AB, UK
{F.Rezwan,Y.2.Sun,N.Davey,R.G.Adams}@herts.ac.uk
[2] Wellcome Trust Sanger Institute, Wellcome Trust Genome Campus, Hinxton,
Cambridge CB10 1SA, UK
ar12@sanger.ac.uk
[3] Benaroya Research Institute at Virginia Mason, 1201 9th Avenue Seattle, WA
98101, USA
mrobinson@benaroyaresearch.org

Abstract. It is known that much of the genetic change underlying morphological evolution takes place in *cis*-regulatory regions, rather than in the coding regions of genes. Identifying these sites in a genome is a non-trivial problem. Experimental methods for finding binding sites exist with some limitations regarding their applicability, accuracy, availability or cost. On the other hand predicting algorithms perform rather poorly. The aim of this research is to develop and improve computational approaches for the prediction of transcription factor binding sites (TFBSs) by integrating the results of computational algorithms and other sources of complementary biological evidence, with particular emphasis on the use of the Support Vector Machine (SVM). Data from two organisms, yeast and mouse, were used in this study. The initial results were not particularly encouraging, as still giving predictions of low quality. However, when the vectors labelled as non-binding sites in the training set were replaced by randomised training vectors, a significant improvement in performance was observed. This gave substantial improvement over the yeast genome and even greater improvement for the mouse data. In fact the resulting classifier was finding over 80% of the binding sites in the test set and moreover 80% of the predictions were correct.

1 Introduction

Deciphering the non-coding regions in a genome is significant not only to understand their functional association with gene coding sequences, but also for discovering the regulatory instructions they specify. Hence studying gene regulation has been a major focus of studies in the fields of biology and bioinformatics. Over recent years the opportunities to study gene regulation has increased markedly

M.A. Lones et al. (Eds.): IPCAT 2012, LNCS 7223, pp. 225–237, 2012.
© Springer-Verlag Berlin Heidelberg 2012

as, with the advent of high-throughput experiments from next-generation sequencing technologies, there is now an unprecedented abundance of genomic data available from a large number of publicly accessible databases. Specifically, non-coding regions in the vicinity of genes may contain short stretches of DNA sub-sequences to which proteins can bind. These regions are known as *cis*-regulatory binding sites or transcription factor binding sites (TFBSs) and are known to finely regulate gene expression [1, 2]. The composition and number of *cis*-regulatory binding sites across multiple non-coding regions give rise to a complex set of gene regulatory networks that encode the regulatory program of a cell. There are very few biological processes that are not influenced by regulatory mechanisms.

In our earlier works [3–8], results of a group of different predictor algorithms were used to produce a combined prediction that is better than any of the individual algorithms. However, these improvements were not very large. The results showed that this approach was still generating lots of incorrect predictions if we compare them to the original annotated binding sites (see Fig. 1). One of the major problems with training a classifier on these combined predictions is that the data constituting the training set can be unreliable. In the datasets used here negative examples were originally just the promoter regions that were not annotated as TFBSs (referred to as promoter negative examples). We were interested in whether the individual prediction algorithm could be misleading when taken from the negatively labeled promoter area. This issue will be addressed in this paper by replacing these promoter negative examples with the ones produced by a randomisation process (randomised negative examples) and the presented results will show a major improvement on the original algorithms.

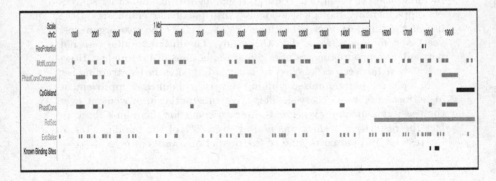

Fig. 1. Computational algorithms produce a lot of false predictions. Here, at each position in the genome sequence we have an annotation and seven algorithmic predictions for the mouse data.

2   Background

There are many reasons why binding site prediction is a difficult problem to solve computationally [10]. The binding sites can be very short and of variable sizes. Furthermore a typical transcription factor can bind to a number of different DNA sequences, albeit with very similar DNA composition. These are just some of the reasons why identifying binding sites computationally is a non-trivial problem rather than a simple pattern recognition or regular expression problem.

There are many experimental and computational approaches for identifying regulatory sites. Many experimental techniques have limitations regarding their applicability, accuracy, availability and cost and therefore are not amenable to a genome-wide approach [9]. Experimental approaches that are genome-wide, such as ChIP-chip and ChIP-seq, are themselves dependent on the availability of specific antibodies and still require additional verification. From a computational standpoint, there are a number of algorithmic strategies for computationally predicting the location of TFBSs by searching matches to a motif model [11], exploiting statistical characteristics of sequence features, clustering genes that share similar patterns of expression for a given biological condition, or measuring evolutionary conservation of DNA sequences to infer TFBSs. Each of these strategies has their own limitations giving rise to many false positives predictions, which can significantly limit their utility. However, there is still scope for improvement by reducing their weaknesses and combining their strengths together. For example, scanning algorithms can only predict those sites that match to the known set of motifs they are supplied with, whereas co-regulatory algorithms can only predict shared, and not unique, binding sites in the regulatory sequences for a set of co-expressed genes. Hence the different subsets of accurately predicted binding sites are complementary and can provide more accurate and reliable predictions if combined together [9].

In our approach, algorithms and supporting biological data were classified using Support Vector Machines (SVM) [12]. Data sets from two organisms were used here - the relatively simple genome of yeast and the more complex genome of the mouse. As we already mentioned that the predictions from the original algorithms in the non-binding sites in the promoter region could be unreliable and this unreliability can be dealt with by introducing the concept of *synthetic negative examples*. This paper will detail the results produced by this approach.

3   Description of Data

3.1   Genomic Data

As mentioned earlier, two datasets from different organisms were used in the experiments described in this paper. The yeast data consisted of 112 annotated promoter sequences, which were selected for training and testing the algorithms, a total of $67,782$ base pairs (bps) of sequence data from the *S. cerevisiae Promoter Database* (SCPD) [13]. For each promoter, 500 bps of sequence were taken immediately upstream from the transcriptional start site were considered sufficient to typically allow full regulatory characterisation in yeast [13]. In cases where annotated binding sites lay outside of this range, then the range was expanded accordingly. Likewise, where a 500 bps upstream region would overlap a coding region then it was truncated accordingly.

Table 1. A summary of yeast and mouse datasets

	Yeast dataset	Mouse dataset
Total number of sequences	112	47
Total sequence length	67,782 bps	60,851 bps
Average sequence length	605 bps	1294.70 bps
Average number of TFBSs per sequence	3.6	2.87
Average TFBSs width	13.2 bps	12.78 bps
Total number of TFBS	400	135
TFBS density in total data set	7.8%	2.85%

Forty seven (47) annotated promoters sequences had been taken with TBFSs for the mouse data and merged together into a single data set from the *ABS* [14] and *ORegAnno* databases [15]. The sequence length, extracted from *ABS*, was typically 500 bps and those from *ORegAnno* were around 2000 bps in length. Most of the promoters were upstream of their associated gene and a few of them were extended over the first exon including intronic regions. Therefore, there are 60851 nucleotides in total in the data set. See Table 1 for details.

We used 12 different prediction algorithms (discussed in [6, 8]) for the yeast data. Combined together they make a single data set, a 67782 by 13 data matrix where the first column is the annotation label and rest of the columns represent scores from the seven prediction algorithms. The label is "1" if it is a part of a TFBS, otherwise "0". In case of the mouse data, seven different prediction algorithms had been used (discussed in [8]) and a 60851 by 8 data matrix formed. Hence default parameter values taken from the literature were used in this study. These parameter values are therefore already selected to be good values in the literature.

3.2   Randomised Negative Example Datasets

To produce the randomised negative examples, all the data vectors labeled as non-binding sites, had been placed into a matrix with a column for each base algorithm. Each column was then independently randomly reordered. This randomised each column vector but maintained the overall statistical properties of each algorithm since all the original algorithm values were still there.

4   Methods

In the method described in this paper, promoter negative examples have been either used or replaced with randomised negative examples in the data sets. Therefore, we have run two types of experiments for both yeast and mouse datasets:

Experiment 1: Using promoter negative examples not annotated as TFBSs from the original data

Experiment 2: Replacing negative examples with randomised negative examples

In addition, we have applied some pre-processing (data division, normalisation and sampling) on the training set and some post-processing on the prediction set. Fig. 2 shows the complete workflow of our method.

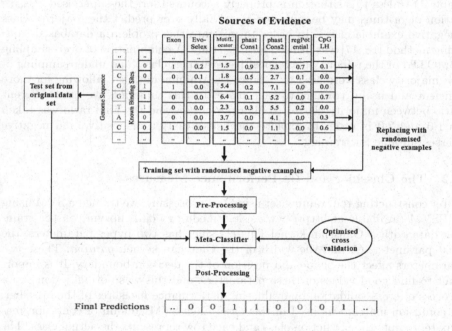

Fig. 2. A complete workflow of the integration of sources of evidence. We train an SVM on combined predictions to produce improved predictions.

4.1   Pre-processing (Preparing Training and Test Sets)

In pre-processing, first we normalised each column and this had been done by subtracting the population mean of each algorithm from an individual score of the algorithm and then dividing the difference by the population standard deviation of the that particular algorithm, in other words each feature was turned into *Z-score*. In our data sets there are also a number of vectors that are repeated. There are repeats that belong to both classes (positive and negative classes). We call them inconsistent vectors. There are also repeats that occur in only one class and these are simply called repeats.Therefore, we searched for any repetitive or inconsistent data vector and eliminated them from the training set. After removing the repetition and inconsistency in data sets, we shuffled the data rows to mix the positive and negative examples randomly and thus we got a new data set. In case of using randomised negative examples, we replaced the promoter negative examples before starting the pre-processing. We took two-third of the new data set as the training data. However we took the test set from the original dataset prior to the pre-processing, which is biologically meaningful since it is a contiguous section from the original genome.

The dataset used for this study is highly imbalanced. By imbalanced, it is meant that the number of positive examples (sites containing TBFSs) is much less than that of negative examples (sites which do not contain TFBSs) (see Table 1).Unless this situation is properly accounted for, the supervised classification algorithms may be expected to trivially over predict the majority class (negative example class). In order to mitigate this problem a databased sampling method [16, 17] was utilised for this study. A combination of over-sampling (SMOTE) of the minority class (positive examples class) and under-sampling of the majority class were used to balance the training data set, allowing for more efficient and useful training to take place. In this sampling, we used different ratio between majority classes to minority classes and the final ratio was taken as 1.0 and 2.0 in order to explore the better margin of positive and negative classes suitable for training.

4.2   The Classifier and Its Performance Measures

After constructing the training set using pre-processing, we trained a SVM using LIBSVM (available at http://www.csie.ntu.edu.tw/ cjlin/libsvm) on the training data with a Gaussian kernel.This classifier has two hyper parameters, the *cost* parameter, C, and the width of the Gaussian kernel, *gamma*. These two parameters affect the shape and position of the decision boundary. It is important to find good values of these parameters, and this was normally done by a process of cross-validation by evaluating performance measures of the classifier. A confusion matrix was used in this case. Each SVM column of confusion matrix represents the prediction class and each row represents the actual class. The confusion matrix has the following entries:

Table 2. A confusion matrix

	Predicted Negatives	**Predicted Positives**
Actual Negatives	True Negatives(TN)	False Positives(FP)
Actual Positives	False Negatives(FN)	True Positives(TP)

The performance measures of a confusion matrix are described by Equations (1) to (5).

$$Accuracy = \frac{TP + TN}{TP + FN + FP + TN} \tag{1}$$

$$Recall = \frac{TP}{TP + FN} \tag{2}$$

$$Precision = \frac{TP}{TP + FP} \tag{3}$$

$$F\text{-}score = \frac{2 \times Recall \times Precision}{Recall + Precision} \tag{4}$$

$$FP\text{-}rate = \frac{FP}{FP + TN} \tag{5}$$

Accuracy (correct classification) can be an ideal performance measure to report the efficiency of a classifier. But as we were dealing with imbalanced data sets, simply using *Accuracy* as the performance measure might not be appropriate. Predicting everything as not belonging to a binding site can give a very good *Accuracy* rate. *Recall* (True positive rate) is the proportion of positive cases that were correctly identified. But *Recall* can easily be increased by over-prediction. One the other hand, *Precision* is the proportion of the predicted positive cases that were correct. Increase in *Precision* can improve the prediction result considerably, but it may decrease the True Positives in the prediction. Combining both *Recall* and *Precision* should be a solution. Hence, *F-score* ought to be a useful performance measure. In addition reducing the *FP-rate* should also be another major concern to verify a classifier's performance.

4.3   Cross-Validation

A modified cross-validation method had been devised in which the model is validated in exactly the same way, as it would be tested. The model was validated with non-sampled validation sets and short predictions were removed. In this new modified version, the data is divided in five subsets for 5-fold cross validation and four subsets are taken as training during cross-validation and the validation set is reconstructed from the corresponding original data sets. The training set in the cross-validation undergoes the same pre-processing to make it balanced. Performance has been measured on the validation data by using *F-score*. The step-by-step description of exactly what we did is shown in Fig. 3.

1. Replace negative examples in the original data set with randomised
negative examples
2. Remove repeats and inconsistent vectors
3. Split data into 3 equal subsets. Take two of the sets as training and
one set from the original data as test set
4. Split the training data into 5 partitions
5. This gives 5 different training (four-fifth) and validation
(one-fifth) sets. The validation set is drawn from the related original
data set
6. Use sampling to produce more balanced training sets
7. For each pair of $C/gamma$ values
   7.1. For each of the 5 training sets
      7.1.1. Train an SVM
      7.1.2. Measure performance on the corresponding validation
      set, exactly as the final test will be measured. So use the
      Performance Measure, after the predictions on the validation
      set have been filtered (by post-processing described in
      Section 4.4)
   7.2. Average the Performance Measure over the 5 trials
8. Choose the $C/gamma$ pair with the best average Performance Measure
9. Pre-process the complete training set and train an SVM with the best
$C/gamma$ combination
10. Test the trained model on the unseen test set

Fig. 3. Pseudocode of the cross-validation method

4.4   Post-processing (Filtering Short Predictions)

The original biological algorithms predict contiguous sets of base pairs as binding
sites. However in the classification approach undertaken here, each base pair
was predicted independently of their neighbouring nucleotides. As a result, the
classifier might output many short predictions, sometimes even with a length of
only one or two. From both yeast and mouse data sets, it could be seen that
the shortest binding site was 5 bps in length and the longest one was more or
less 13 bps. Therefore, predictions with a length equal or less than a threshold
value had been removed (replaced the positive prediction with a negative one)
and then the effect of the performance was measured. In this study, different
threshold values (from 4 bps to 7 bps) had been used to explore possible feasible
threshold sizes.

5   Results and Discussion

5.1   Yeast Dataset

Before presenting our experimental results, let us see how the base algorithms
perform for identifying *cis*-binding sites on the same test set we used in all the
experiments described in Section 4. We calculated the performance measures of

the 12 algorithms discussed in [6, 8] and among them we took the best result from the algorithm, *Fuzznuc*. The confusion matrix of the best algorithm, (*Fuzznuc*), is shown in Table 3.

Table 3. Confusion matrix of *Fuzznuc* on the yeast data

	Predicted Negatives	Predicted Positives
Actual Negatives	TN = 83%	FP = 10%
Actual Positives	FN = 4%	TP = 3%

From the results it is evident that *Fuzznuc* incorrectly predicted a relatively high number of sites as binding sites. So in this case the False Positives are more than three times greater than the True Positives, which even makes the best algorithm unreliable.

Table 4. The results of *Fuzznuc*, yeast data with the original negative examples (*Experiment 1*) and yeast data with randomised negative examples (*Experiment 2*)

	TP	FP	FN	TN	Recall	Precision	F-score	FP-rate
Fuzznuc	635	2227	953	18779	0.4	0.22	0.25	0.11
Yeast Experiment 1	643	1409	2165	18377	0.23	0.31	0.27	0.07
Yeast Experiment 2	1748	67	1061	19718	0.62	0.96	0.76	0.003

Table 4 presents the results using the trained SVM on the original yeast data and the yeast data with randomised negative examples as described in Section 4. For comparison purposes, the result of *Fuzznuc* on the original data is also being presented. It is apparent that the original combined predictor is only marginally better than the best base algorithm, *Fuzznuc*. In fact the *F-score* only improves from 0.25 to 0.27. The number of False Positive decreases and number of False Negative increases. However, the introduction of randomised negative examples has a substantial impact and we can see that using synthetic negative examples gives almost a three-fold increase in the True Positives and dramatically decreases the number of False Positive predictions, hence yielding very high precision, 96%. We therefore have actually produced a reasonably accurate predictor for identifying *cis*-binding sites for the yeast.

5.2   Mouse Dataset

Now let us see how the base algorithms perform for identifying *cis*-binding sites on the test set we used in the mouse experiment. We calculated the performance measures of the seven algorithms discussed in [8] and amongst these the best performing algorithm is *EvoSelex*. Its confusion matrix of the best algorithm (*EvoSelex*) is shown in Table 5.

Table 5. Confusion matrix of *EvoSelex* on the mouse data

	Predicted Negatives	Predicted Positives
Actual Negatives	TN = 79%	FP = 16.5%
Actual Positives	FN = 3%	TP = 1.5%

From these results in Table 5, it can be observed that the quality of the predicted binding sites is very poor. More than 90% of the predicted binding sites are actually not binding sites, therefore Precision is only 8%. As a result the *F-score* is also not high. This happens because the algorithm tries to annotate many non-binding sites as binding sites and this leads to a high *Recall* but leaves the *Precision* low.

Table 6. The results of *EvoSelex*, mouse data with the original negative examples (*Experiment 1*) and mouse data using randomised negative examples (*Experiment 2*)

	TP	FP	FN	TN	Recall	Precision	F-score	FP-rate
EvoSelex	273	3139	511	14985	0.35	0.08	0.13	0.17
Mouse Experiment 1	191	908	593	17216	0.24	0.17	0.20	0.05
Mouse Experiment 2	594	0	190	18124	0.76	1.0	0.86	0.00

Table 6 presents the results using the trained SVM on the original mouse data and the mouse data using randomised negative examples as described in Section 4. For comparison purposes, the result of *EvoSelex* on the original data is also being presented. It is apparent that the original combined predictor is only marginally better than the best base algorithm, *EvoSelex*. In fact the number of false positive decreases and number of false negative increases, *Precision* has more than doubled and the *F-score* improves from 0.13 to 0.20. However, once again using synthetic negative data produces a trained model that is of a much higher quality than the model derived from the original data, the *F-score* jumped from 0.20 to 0.86. Examining rows three and four of Table 6 shows that number of True Positives has been more than trebled whilst, remarkably, the False Positive predictions have completely disappeared. Therefore, here our trained classifier is able to predict the presence of binding sites with a high degree of confidence.

5.3   Visualisation of the Predictions

As well as assessing predictions based on performance measures, we produced a visualisation of the predictions on the mouse data set. Because of the results from the mouse genome is so good, we decided to visualise the predictions over a subset of the mouse data. Three subsequences of the mouse genome are visualised. The regions are 2000 bps upstream of the gene *MyoD1*, 500 bps upstream of the gene *Vim*, and 2000 bps upstream of the gene *U36283*.

In Fig. 4, the upper seven results are from the original prediction algorithms. The next two results are the prediction results from the two different types of

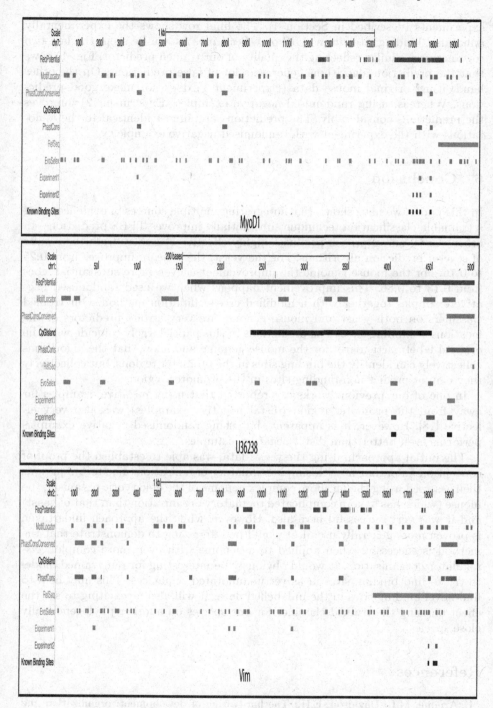

Fig. 4. Visualisation of the predictions on the mouse data

experiments (described in Section 4). The final row shows the experimentally annotated binding sites from *ABS* or *ORegAnno*. As can be seen, the last two rows are very similar reflecting the quality of our trained predictor. Fig. 4 shows that the prediction algorithms generate a lot of false predictions. On the other hand, using original mouse data (Experiment 1) does not make good predictions. Whereas, using randomised negative examples (Experiment 2) improves the predictions considerably. The predictions are almost identical to the annotations with the experiment with randomised negative example.

6 Conclusion

In this paper, we have shown that integrating multiple sources of evidence using a trainable classification technique substantially improves TFBS predictions. In other words, our approach produces highly effective classifiers from the results of several prediction algorithms. For the yeast, the *F-score* improves from 0.25 to 0.76. For the mouse genome the improvement is even more substantial, goes from 0.13 to 0.86. This improvement happens when we used randomised negative examples together with a modified cross-validation method. Our trained classifiers on both yeast and mouse genome are very strong predictors of the locations of binding sites. The predictions in this paper largely coincide with the original label particularly for the mouse genome and show that the algorithms collectively can identify the binding sites in the promoter regions, but collectively they cannot predict non-binding sites in the promoter regions.

In one of our previous works, we reported that using negative examples far away from the promoter region (distal negative examples) was also very effective [18]. However, it is apparent that using randomised negative examples performs even better than distal negative examples.

The initial approach, using the yeast data, was able to establish the proof of concept on an organism with simple regulatory organisation. The technique was then assessed using the far more complex genome of the eukaryotic organism, the mouse (which has a more complicated regulatory organisation than that of yeast) and it was very successful here also. However, while the approach undertaken is proven more generally useful, it would be interesting to demonstrate that the method is successful when applied to a genome with even more complex *cis*-regulatory organisation. It would obviously be interesting for our trained model to try to find binding sites in as yet unannotated sequences. This may help to find novel binding sites in the unlabelled data. It will also be exciting to see the effect of using the trained classifier for one species on another phylogenetically close species.

References

1. Arnone, M.I., Davidson, E.H.: The hardwiring of development: organization and function of genomic regulatory systems. Development 124, 1851–1864 (1997)

2. Davidson, E.H.: Genomic Regulatory Systems: Development and Evolution. Academic Press (2001)
3. Sun, Y., Robinson, M., Adams, R., Davey, N., Rust, A.G.: Predicting Binding Sites in the Mouse Genome. In: ICMLA, pp. 476–481. IEEE Computer Society (2007)
4. Sun, Y., Robinson, M., Adams, R., Rust, A.G., Davey, N.: Prediction of Binding Sites in the Mouse Genome Using Support Vector Machines. In: Kůrková, V., Neruda, R., Koutník, J. (eds.) ICANN 2008, Part II. LNCS, vol. 5164, pp. 91–100. Springer, Heidelberg (2008)
5. Sun, Y., Robinson, M., Adams, R., te Boekhorst, R., Rust, A.G., Davey, N.: Integrating genomic binding site predictions using real-valued meta-classiers. Neural Comput. Appl. 18, 577–590 (2009)
6. Sun, Y., Castellano, C.G., Robinson, M., Adams, R., Rust, A.G., Davey, N.: Using pre and post-processing methods to improve binding site predictions. Pattern Recogn. 42, 1949–1958 (2009)
7. Robinson, M., Castellano, C.G., Adams, R., Davey, N., Sun, Y.: Identifying Binding Sites in Sequential Genomic Data. In: de Sá, J.M., Alexandre, L.A., Duch, W., Mandic, D.P. (eds.) ICANN 2007, Part II. LNCS, vol. 4669, pp. 100–109. Springer, Heidelberg (2007)
8. Robinson, M., Castellano, C.G., Rezwan, F., Adams, R., Davey, N., Rust, A., Sun, Y.: Combining experts in order to identify binding sites in yeast and mouse genomic data. Neural Networks 21(6), 856–861 (2008)
9. Tompa, M., Li, N., Bailey, T.L., Church, G.M., De Moor, B., Eskin, E., Favorov, A.V., Frith, M.C., Fu, Y., Kent, W.J., Makeev, V.J., Mironov, A.A., Noble, W.S., Pavesi, G., Pesole, G., Régnier, M., Simonis, N., Sinha, S., Thijs, G., van Helden, J., Vandenbogaert, M., Weng, Z., Workman, C., Ye, C., Zhu, Z.: Assessing computational tools for the discovery of transcription factor binding sites. Nat. Biotechnol. 23(1), 137–144 (2005)
10. Brown, C.T.: Computational approaches to finding and analyzing cis-regulatory elements. Methods Cell Biol. 87, 337–365 (2008)
11. Stormo, G.D.: DNA binding sites: representation and discovery. Bioinformatics 16, 16–23 (2000)
12. Cortes, C., Vapnik, V.: Support-Vector Networks. Machine Learning 20 (1995)
13. Zhu, J., Zhang, M.Q.: SCPD: a promoter database of the yeast Saccharomyces cerevisiae. Bioinformatics 15, 607–611 (1999)
14. Blanco, E., Farré, D., Albà, M.M., Messeguer, X., Guigó, R.: ABS: a database of Annotated regulatory Binding Sites from orthologous promoters. Nucleic Acids Res. 34(Database issue), D63–D67 (2006)
15. Montgomery, S.B., Griffith, O.L., Sleumer, M.C., Bergman, C.M., Bilenky, M., Pleasance, E.D., Prychyna, Y., Zhang, X., Jones, S.J.M.: ORegAnno: An open access database and curation system for literature-derived promoters, transcription factor binding sites and regulatory variation. Bioinformatics (March 2006)
16. Chawla, N.V., Bowyer, K.W., Hall, L.O., Kegelmeyer, W.P.: SMOTE: Synthetic Minority Over-sampling Technique. J. Artif. Intell. Res. (JAIR) 16, 321–357 (2000)
17. Radivojac, P., Chawla, N.V., Dunker, A.K., Obradovic, Z.: Classification and knowledge discovery in protein databases. J. Biomed. Inform. 37, 224–239 (2004)
18. Rezwan, F., Sun, Y., Davey, N., Adams, R., Rust, A.G., Robinson, M.: Effect of Using Varying Negative Examples in Transcription Factor Binding Site Predictions. In: Giacobini, M. (ed.) EvoBIO 2011. LNCS, vol. 6623, pp. 1–12. Springer, Heidelberg (2011)

Finding the Minimal Gene Regulatory Function in the Presence of Undefined Transitional States Using a Genetic Algorithm

Rocio Chavez-Alvarez, Arturo Chavoya*, and Cuauhtemoc Lopez-Martin

Department of Information Systems, Universidad de Guadalajara,
Periferico Norte 799-L308, Zapopan, Jal., Mexico 45100
{rociochavezmx,achavoya,cuauhtemoc}@cucea.udg.mx
http://www.cucea.udg.mx

Abstract. After the sequencing of whole genomes and the identification of the genes contained in them, one of the main challenges remaining is to understand the mechanisms that regulate the expression of genes within the genome in order to gain knowledge about structural, biochemical, physiological and behavioral characteristics of organisms. Some of these mechanisms are controlled by so-called Genetic Regulatory Networks (GRNs). Boolean networks can help model biological GRNs. In this paper, a genetic algorithm is used to make inferences in Boolean networks, in combination with the Quine-McCluskey algorithm, when not all the output states of the genes have been determined. This lack of information could be treated as "don't care" states. Genetic algorithms are useful in multi-objective optimization problems, such as minimization of Gene Regulatory Functions, where it is important not only to have the smallest quantity of disjunctions, but also the smallest quantity of genes involved in the regulation.

Keywords: Genetic regulatory networks, Boolean networks, Don't care states, Genetic algorithm, Quine-McCluskey algorithm.

1 Introduction

In order to discover the mechanisms that regulate gene expression in Genetic Regulatory Networks (GRNs), it is necessary to know how genes are interconnected and the kind of influence they have on each other. In GRNs, genes can be translated into proteins through messenger RNA (mRNA) molecules. These proteins can activate or inhibit the expression of other genes by binding to regulatory regions present near the genes [1]. This means that the expression of regulated genes can be detected by the existence or the absence of the mRNA produced by regulatory genes. This mRNA can be quantified by oligonucleotide microarray chips. These chips are able to generate large amounts of data stored in databases containing the quantities of mRNA strands that are being produced simultaneously by thousands of genes [2]. An example of these databases

---

* Corresponding author.

M.A. Lones et al. (Eds.): IPCAT 2012, LNCS 7223, pp. 238–249, 2012.

is the one provided by Cho and Campbell [3]. This database contains the expression of 6600 genes obtained in 17 time points during the cell cycle of the yeast *Saccharomyces cerevisiae*.

Data obtained from gene expression can be transformed into binary values in such a way that gene expression is represented by 1 when the gene is being activated and 0 when it is inhibited [4][5]. Once gene expression in GRNs has been transformed to binary values, it can then be analyzed as a Boolean network (BN).

Boolean networks were first proposed by Kauffman to model GRNs [6]. This approach has been useful to describe real gene regulatory relations such as the *Drosophila* segment polarity network, whereas attractors in BNs have been helpful to study cellular phenotypes in living organisms [7]. Furthermore, BNs have several advantages such as the ease of their interpretation and their dynamic nature, which make them suitable to simulate gene regulatory events [8].

A Boolean Network is a directed graph $G = \{V, F\}$ that is composed of a set of nodes $V = (g_1, \ldots, g_n)$ that represent genes and a set of Boolean functions $F = (f_1, \ldots, f_n)$ that describe activation or repression of genes. Each Boolean function is composed of k genes that regulate the expression of the target gene.

In Boolean networks, inference can be done through the use of the Quine-McCluskey algorithm, as described in [9]. This algorithm, developed by Willard Van Orman Quine and Edward J. McCluskey [10][11][12], allows the simplification of Boolean functions and it could be useful in discovering gene modularity, which can be described as a sum of products. For example, the function $g_a = g_b g_c + g_d$ means that in order to activate g_a, the combination of proteins produced by g_b and g_c or the presence of the protein produced by g_d is necessary. A more detailed explanation of the Quine-McCluskey algorithm is given in Sect. 2.

The Quine-McCluskey algorithm could also be useful in the detection of gene feedback, as accomplished by the REVEAL algorithm [13]. Section 3 shows how similar results were obtained with both algorithms.

In order to apply the Quine McCluskey algorithm, it is necessary to have 2^N states, where N is the number of genes in the network. Frequently, it is not possible to have such a quantity of states due to the difficulty and the cost of the tests performed to measure gene expression (i.e. the quantity of mRNA produced by the genes). Only a few tests are usually performed, meaning that the transition state table for the network to be inferred is typically not fully defined. Undefined states can be represented by so-called "don't care" states, which are symbolized by an "X" in the transition state table (see an example in Table 10). In the presence of "don't care" states, genetic algorithms could help find the minimal logical function of regulated genes. An explanation of objective function optimization with genetic algorithms can be found in Sect. 4, whereas results from the application of the genetic algorithm is presented in Sect. 5. Finally, Sect. 6 presents the conclusions and the future work.

2 Boolean Function Simplification Using the Quine-McCluskey Algorithm

In order to explain how the Quine-McCluskey algorithm works, a simple example is presented next. Suppose we have the observations presented in Table 1 of the behavior of five genes (termed from g_1 to g_5) and we wanted to know the way in which g_1, g_2, g_3 and g_4 have an influence over g_5. As described above, value 1 represents an active gene and value 0 represents an inhibited gene.

Table 1. Observations made on five genes

g_1	g_2	g_3	g_4	g_5
0	0	1	0	0
0	0	0	0	1
0	0	1	1	1
0	0	0	1	1
0	1	0	1	1
0	1	1	1	1
0	1	1	0	0
1	1	1	1	0
1	0	1	1	1
0	1	0	0	1
1	1	1	0	0
1	0	0	0	0
1	0	0	1	1
1	0	1	0	0
1	1	0	0	0
1	1	0	1	0

Step 1. To apply the Quine-McCluskey algorithm, the first step is to place the gene input observations in an ascending binary order, from 0000 to 1111, where 0000 is equivalent to 0 in the decimal system and 1111 represents 15. With this information we can build the truth table for g_5 presented in Table 2.

In the truth table, all the output states for g_5 are defined, i.e. every state in the table has a value of either 0 or 1.

Step 2. Group in a table all the combinations of $g_1g_2g_3g_4$ for which g_5 has value 1, as in Table 3. The value of 1 means that g_5 is being activated when the listed combinations are present.

Step 3. All these combinations of $g_1g_2g_3g_4$ are converted from binary to decimal and the number of 1s are counted and grouped as shown in Table 4.

Step 4. Pair terms that have only one difference between them and count again the number of 1s. Replace the bit that differs with a "-". For example, for pair $(0, 1)$, all the bits are equal, except for the one that corresponds to gene g_4. Thus, the new value for $g_1g_2g_3g_4$ is 000-.

Table 2. Truth table for g_5

g_1	g_2	g_3	g_4	g_5
0	0	0	0	1
0	0	0	1	1
0	0	1	0	0
0	0	1	1	1
0	1	0	0	1
0	1	0	1	1
0	1	1	0	0
0	1	1	1	1
1	0	0	0	0
1	0	0	1	1
1	0	1	0	0
1	0	1	1	1
1	1	0	0	0
1	1	0	1	0
1	1	1	0	0
1	1	1	1	0

Table 3. Combinations of genes $g_1 g_2 g_3 g_4$ that activate g_5

g_1	g_2	g_3	g_4	g_5
0	0	0	0	1
0	0	0	1	1
0	0	1	1	1
0	1	0	0	1
0	1	0	1	1
0	1	1	1	1
1	0	0	1	1
1	0	1	1	1

Table 4. Conversion from binary to decimal and counting of 1s in binary numbers

Decimal	g_1	g_2	g_3	g_4	Quantity of 1s
0	0	0	0	0	0
1	0	0	0	1	1
4	0	1	0	0	
3	0	0	1	1	2
5	0	1	0	1	
9	1	0	0	1	
7	0	1	1	1	3
11	1	0	1	1	

If there is one term that cannot be grouped, it is reserved to be part of the final solution.

Table 5. Paired terms and counting of 1s

Pair of Terms	Comparison	Quantity of 1s
(0,1)	0 0 0 -	0
(0,4)	0 - 0 0	
(1,3)	0 0 - 1	1
(1,5)	0 - 0 1	
(1,9)	- 0 0 1	
(4,5)	0 1 0 -	
(3,7)	0 - 1 1	2
(3,11)	- 0 1 1	
(5,7)	0 1 - 1	
(9,11)	1 0 - 1	

Repeat step 4 until it is not possible to find pairs of terms. The final reduction results are presented in Table 6.

Table 6. Final reduction results

Pair of Terms	Comparison	Quantity of 1s
(0,1,4,5)	0 - 0 -	0
(0,4,1,5)	0 - 0 -	
(1,3,5,7)	- 0 - 1	1
(1,3,9,11)	- 0 - 1	
(1,5,3,7)	0 - - 1	
(1,9,3,11)	- 0 - 1	

As there is not a further possible reduction, the compared terms left are the solution and the corresponding logic terms are as shown in Table 7.

Interpretation. The function that describes g_5 is

$$g_5 = \bar{g}_1\bar{g}_3 + \bar{g}_2 g_4 + \bar{g}_1 g_4 \,, \tag{1}$$

where "+" means logical "OR", $\bar{g}_1\bar{g}_3$ is the negation of g_1 and of g_3, which means the absence of g_1 AND the absence of g_3, $\bar{g}_2 g_4$ means the absence of g_2 AND the presence of g_4, and finally $\bar{g}_1 g_4$ means the absence of g_1 AND the presence of g_4.

To solve this example in MATLAB, the syntax for the Quine-McCluskey algorithm is

Table 7. Final result of the application of the Quine-McCluskey algorithm

g_1 g_2 g_3 g_4	Logic terms
0 - 0 -	$\bar{g}_1\bar{g}_3$
- 0 - 1	$\bar{g}_2 g_4$
0 - - 1	$\bar{g}_1 g_4$

$$R = minBool([0\ 1\ 4\ 3\ 5\ 9\ 7\ 11])\ . \tag{2}$$

The *minBool* function is used with all the decimal numbers corresponding to combinations for which g_5 has value 1 as parameter. It is not necessary to place the numbers in order, as the Quine-McCluskey algorithm is able to order them.

The solution is shown as

$$
\begin{array}{cccc}
1 & 2 & 3 & 4 \\
\hline
-1 & 0 & -1 & 0 \\
0 & -1 & 0 & 1 \\
-1 & 0 & 0 & 1
\end{array}
$$

where numbers 1, 2, 3 and 4 correspond to genes g_1, g_2, g_3 and g_4, respectively.

The value of zero in the result represents that it is an irrelevant variable, -1 represents the negation of the variable and 1 represents the positive value of the variable.

All the variables in a row are joined by *AND* operators and all rows are connected by *OR* operators. As can be seen, the solution is the same as the one presented in (1).

3   Gene Feedback Detection

As mentioned in Sect. 1, we found that the Quine-McCluskey algorithm can detect gene feedback in time-series in the same manner as the REVEAL algorithm [13]. This can be achieved by taking the input states from time t and the output states from time $t + 1$.

The REVEAL algorithm was developed by Liang et al. [13]. They used Shannon entropy to make inferences for finding GRNs. Shannon entropy is the probability of observing a particular event or symbol within a given sequence [14]. To test their model, they built a Boolean network with several attractors (i.e. repeating state cycles). Table 8 contains the complete state transition table that defines the network dynamics.

The results that Liang et al. [13] obtained were: Gene A is regulated by gene B, gene B is activated in the presence of either gene A or gene C, and gene C is activated for the combination of gene A and gene B or the combination of gene B and gene C or the combination of gene A and gene C. In symbolic form:

$$A_{(t+1)} = B\ , \tag{3}$$

Table 8. State Transition Table from [13]

A B C	A_{t+1}	B_{t+1}	C_{t+1}
0 0 0	0	0	0
0 0 1	0	1	0
0 1 0	1	0	0
0 1 1	1	1	1
1 0 0	0	1	0
1 0 1	0	1	1
1 1 0	1	1	1
1 1 1	1	1	1

$$B_{(t+1)} = A \text{ or } C, \tag{4}$$

$$C_{(t+1)} = (A \text{ and } B) \text{ or } (B \text{ and } C) \text{ or } (A \text{ and } C). \tag{5}$$

As can be seen, node C presents feedback, which means that gene C has an influence on itself.

The results we obtained by applying the Quine-McCluskey algorithm as implemented by Mihailov et al. [16] to the same data are presented in Table 9. They are the same as the ones obtained by Liang et al. [13], including the detection of feedback for gene C. Thus, the Quine-McCluskey algorithm can be used not only to find directed acyclic graphs but also cyclic ones.

Table 9. Results obtained using the Quine-McCluskey Algorithm

Evaluated variable	Result
$A_{(t+1)}$	B
$B_{(t+1)}$	A + C
$C_{(t+1)}$	AB + BC + AC

4   Boolean Function Minimization with Multi-objective Genetic Algorithms

Genetic Algorithms (GAs) are search optimization methods that can be used to find a set of parameters that maximize or minimize a given fitness function. They were introduced by John Holland in 1975. GAs work with an initial set of individuals or chromosomes that represent potential solutions to the optimization problem. Each individual is evaluated and ranked according to its fitness value and the best individuals are chosen to be part of the new generation. GAs typically use operators, such as crossover and mutation, to introduce genetic variety to the population.

The optimization process is generally terminated when the desired solution has been found or when a predefined number of generations has been reached.

Due to the stochastic nature of the algorithm, it can happen that the global optimum is not always found [15].

For GRNs, it is desirable to find not just the minimum quantity of terms in the logical function (disjunctions), but also the minimum quantity of genes involved in their regulation. In the presence of "don't care" states, these two objectives can be achieved by the Multi-Objective Genetic Algorithm developed by Popov et al. [15] in MATLAB, which evaluates and optimizes the function obtained through the Quine-McCluskey algorithm. The original goal of this genetic algorithm was the synthesis of finite state machines, which are one way to describe logic circuits. The code is free and it can be found in Andrey Popov's website.

The objective functions considered in the genetic algorithm described in [15] are:

- f_1 : Number of disjunctions
- f_2 : Number of elements in disjunctions
- f_3 : Total number of inverted elements.

The objective function f_3 was introduced in [15] because it is important to have the minimum quantity of logic gates including "NOT" gates in the reduction of logic circuits. The goal of the application of the GA is to find an individual in the population that minimizes these objective functions.

The GA uses 1s, 0s and "don't care" states as input, instead of the decimal numbers used in the Quine-McCluskey algorithm. In order to illustrate the application of the algorithm, an example is presented next.

Consider gene C from Sect. 3 and suppose that the output states for the 000 and 111 input states are unknown (see Table 10).

Table 10. State transition for gene C

A B C	C_{t+1}
0 0 0	X
0 0 1	0
0 1 0	0
0 1 1	1
1 0 0	0
1 0 1	1
1 1 0	1
1 1 1	X

The input for the genetic algorithm would be [B[1] 0 0 1 0 1 1 B[2]]. All the known output states are fixed in their corresponding positions, whereas the unknown states will be replaced by individuals in the genetic algorithm population to find the minimal solution with the Quine-McCluskey algorithm. These unknown states are represented here by B[1] and B[2].

After application of the algorithm, the result obtained was the same as the one presented in Table 9 for C_{t+1}.

5   Application of the Genetic Algorithm to a Melanoma Dataset

Many tests were performed with the GA described above taking different numbers of regulatory genes. We obtained 100 percent accuracy with up to four input genes when a small amount of known states is used. Thus, it was considered desirable to find the smallest predictor set of genes and apply the GA to those genes.

In order to test the proposed GA, the Melanoma dataset used by Xiao and Dougherty [9], and originally obtained from [17] was used. Our goal was to obtain the same results as these authors, but applying this computational intelligence technique. The dataset consists of 31 expression profiles and it considers 10 genes: RET-1, HADHB, MMP-3, S100P, pirin, MART-1, synuclein, STC2, PHO-C and WNT5A, represented from g_1 to g_{10}, respectively, to simplify the notation. For g_{10}, there are two predictor sets because this gene has two different values (1 and 0) for two identical input gene sequences from g_1 to g_9 (see row number one and row number three in Table 11). This means that under two different contexts, the same combination of input genes produce different outputs for g_{10} with a given probability [9].

Table 11. Filtered melanoma states, as presented in [9]

g_1	g_2	g_3	g_4	g_5	g_6	g_7	g_8	g_9	g_{10}	counts
1	0	0	1	1	1	1	1	0	1	5
1	1	0	1	1	1	0	0	0	0	1
1	0	0	1	1	1	1	1	0	0	2
0	1	0	0	1	1	0	0	0	1	6
0	1	1	0	1	0	0	0	1	1	1
1	0	1	1	1	0	1	1	1	0	3
1	0	1	0	0	0	1	1	1	0	10
0	0	1	0	1	1	0	0	0	0	1
0	1	0	1	0	1	0	0	1	0	1
0	0	1	1	0	1	1	0	1	0	1

Using the information contained in Table 11, Xiao and Dougherty applied a "prune and minimize" algorithm to obtain the minimal predictor set for each gene, as shown in Table 12. This predictor set was used in the present work to test the proposed genetic algorithm.

The GA was applied when "don't care" states were present. As an example, consider the truth table shown in Table 13 for g_3 with predictor genes g_1, g_2 and g_6. It should be noted that the combinations of 000 and 110 for these three genes are not present in Table 11. In the cases where all the predictor gene states were well defined for the output genes, the Quine-McCluskey algorithm alone was applied (see Table 14). The number of generations and individuals in the population for the genetic algorithm were 10 and 30, respectively. These

Table 12. Minimal predictor set for each gene, as presented in [9]

Gene number	Predictor set
g_1	g_4, g_5
g_2	g_3, g_6, g_7
g_3	g_1, g_2, g_6
g_4	g_1, g_5
g_5	g_1, g_4
g_6	g_1, g_2, g_3
g_7	g_1, g_2, g_4
g_8	g_1, g_2
g_9	g_5, g_6
$g_{10}(Function1)$	g_2, g_3, g_4
$g_{10}(Function2)$	g_2, g_4

parameters were sufficient to get the minimal logical functions for all the genes. The same results were obtained with a higher number of generations or greater populations; for example, runs with more than 1000 generations were performed obtaining the same results.

Table 13. Truth table for g_3 with predictor genes g_1, g_2 and g_6

g_1	g_2	g_6	g_3
0	0	0	X
0	0	1	1
0	1	0	1
0	1	1	0
1	0	0	1
1	0	1	0
1	1	0	X
1	1	1	0

The combination of the GA and the Quine-McCluskey algorithm as applied in the present work was able to generate the same results as those obtained by Xiao and Dougherty [9]. Runs were made in a personal computer with the time results presented in Table 14 for each function, showing that this combination can be a suitable alternative for making inferences to find GRNs in the presence of undefined transition states.

Table 14. Results obtained by applying the genetic algorithm and the Quine-McCluskey algorithm. QM = Quine-McCluskey alone; GA = Genetic Algorithm. Time results are the average of 10 runs.

Gene number	Boolean function	Procedure	Time required (seconds)
g_1	$g_4 g_5 + \bar{g}_4 \bar{g}_5$	QM	0.001
g_2	$\bar{g}_3 \bar{g}_7 + \bar{g}_6 \bar{g}_7$	GA	0.368
g_3	$\bar{g}_1 \bar{g}_2 + \bar{g}_6$	GA	0.382
g_4	$g_1 g_5 + \bar{g}_1 \bar{g}_5$	QM	0.001
g_5	$g_1 g_4 + \bar{g}_1 \bar{g}_4$	QM	0.001
g_6	$\bar{g}_1 \bar{g}_2 + \bar{g}_3$	GA	0.451
g_7	$g_1 \bar{g}_2 + \bar{g}_2 g_4$	GA	0.390
g_8	$g_1 \bar{g}_2$	QM	0.005
g_9	$\bar{g}_5 + \bar{g}_6$	QM	0.001
$g_{10(Function1)}$	$\bar{g}_2 \bar{g}_3 + g_2 \bar{g}_4$	GA	0.396
$g_{10(Function2)}$	$g_2 \bar{g}_4$	QM	0.001
		Total time	**1.997**

6 Conclusions and Future Work

In this paper we propose the application of a genetic algorithm to find the minimal logic function in the inference of genetic regulatory networks in combination with the Quine-McCluskey algorithm, which could be useful for finding gene modularity. In the case where defined output states are not available, it is proposed the use of a multi-objective genetic algorithm, which replaces these states by 1s or 0s until finding the minimal combination of disjunction, conjuntions and inverted signals in a short time. This solution, which took approximately 2 seconds to find, is apparently faster than the one proposed by Xiao and Dougherty, which took around 6 seconds; however, these authors did not provide details as to the hardware where they tested their algorithm. This result will be confirmed once we have developed the "prune and minimized" algorithm to include it in the whole process and compare the two approaches in a more direct manner. As an additional result, we found that another advantage of the use of the Quine-McCluskey algorithm is that it can be used to find gene feedback in time-series datasets.

As future work, we intend to perform tests on more biological time-series datasets to define an inference method that can help discover modularity in genetic regulatory networks. We also plan to vary the GA parameters with the goal of reducing optimization times even further.

References

1. Chen, T., Filkov, V., Skiena, S.: Identifying gene regulatory networks from experimental data. In: Proceedings of the Third Annual International Conference on Computational Molecular Biology (RECOMB 1999), pp. 94–103. ACM, New York (1999)
2. De Jong, H.: Modeling and simulation of genetic regulatory systems: a literature review. Journal of Computational Biology 9(1), 67–103 (2002)
3. Cho, R., Campbell, M.A.: Genome-Wide Transcriptional Analysis of the Mitotic Cell Cycle. Molecular Cell 2, 65–73 (1998)
4. Kim, H., Lee, J.K., Park, T.: Boolean networks using the chi-square test for inferring large-scale gene regulatory networks. BMC Bioinformatics 8, 37 (2007)
5. Shmulevich, I.: Binary Analysis and Optimization-Based Normalization of Gene Expression Data. Bioinformatics 18(4), 555–565 (2002)
6. Kauffmann, S.A.: Metabolic Stability and Epigenesis in Randomly Constructed Genetic Nets. J. Theoret. Biol. 22, 437–467 (1969)
7. Xiao, Y.: A Tutorial on Analysis and Simulation of Boolean Gene Regulatory Network Models. Current Genomics 10, 511–525 (2009)
8. Hecker, M., Lambeck, S., Toepfer, S., van Someren, E., Guthke, R.: Gene regulatory network inference: data integration in dynamic models-a review. Biosystems 96(1), 86–103 (2009)
9. Xiao, Y., Dougherty, E.: Optimizing Consistency-Based Design of Context-Sensitive Gene Regulatory Networks. IEEE Transactions on Circuits and Systems 53(11), 2431–2437 (2006)
10. Quine, W.V.: The Problem of Simplifying Truth Functions. Am. Math. Monthly 59, 521 (1952)
11. Quine, W.V.: A Way to Simplify Truth Functions. Am. Math. Monthly 62, 627 (1955)
12. McCluskey, E.J.: Minimization of Boolean Functions. Bell Syst. Tech. J. 35, 1417 (1956)
13. Liang, S.: REVEAL, A General Reverse Engineering Algorithm for Inference of Genetic Network Architectures. In: Pacific Symposium on Biocomputing, vol. 3, pp. 18–29 (1998)
14. Shannon, C.E., Weaver, W.: The mathematical theory of communication. University of Illinois Press (1963)
15. Popov, A., Filipova, K.: Genetic Algorithms Synthesis of Finite State Machines. In: Proceedings of the 27th Spring Seminar on Electronics Technology, pp. 388–392 (2004)
16. Mihailov, S., Popov, A., Filipova, K., Kasev, N.: Comparative Analysis of Boolean Functions Minimization in Terms of Symplifying the Synthesis. In: First International Congress of Mechanical and Electrical Engineering and Technologies, pp. 273–276 (2002)
17. Bittner, M.L., Meltzer, P.: Molecular classification of cutaneous malignant melanoma by gene expression profiling. Nature 406, 536–540 (2000)

Extracting Tailored Protein Complexes
from Protein-Protein Interaction Networks

Hiroshi Okamoto[1,2]

[1] Reserach & Technology Group, Fuji Xerox Co., Ltd.
6-1 Minatomirai, Nishi-ku, Yokohama-shi, Kanagawa 220-8668, Japan
[2] RIKEN Brain Science Institute
2-1 Hirosawa, Wako, Saitama 351-0198, Japan
hiroshi.okamoto@fujixerox.co.jp

Abstract. Suppose that we wish to know a group of proteins responsible for a certain cellular biological process. Here we propose to infer such a protein complex from a protein-protein interaction network by using a class of algorithm, which has originally been developed to achieve web page ranking that reflects user's personal interest or context. The inference of proteins responsible for a given biological process, namely, personalized ranking of proteins is whereby performed in analogy with personalized ranking of web pages. Searching for the best approach to personalized protein ranking, we carry out a series of experiments to compare the performance between two major personalized ranking methods: the personalized PageRank algorithm and the continuous-attractor dynamics algorithm, both applied to a yeast protein-protein interaction network. Results of these comparison experiments suggest that the continuous-attractor dynamics algorithm is the most efficient for personalized protein ranking.

Keywords: Bioinformatics, Interactome, Protein-protein interaction network, Personalized PageRank algorithm, Continuous-attractor dynamics.

1   Introduction

Advances in DNA microarray or yeast two hybridization (Y2H) techniques have enabled us to detect interactions between a large number of genes or proteins at one time. These trends lead us to a perspective of 'transcriptome' or 'interactome', which is embodied by large-scale networks comprising genes or proteins as nodes and interactions between them as links [1, 2].

Until now, identifying proteins responsible for individual cellular biological processes (such as 'regulation of spindle pole body separation' or 'malignant alteration') have required us to carry out separate experiments. However, protein-protein interaction (PPI) networks contain, at least in principle, all proteins and possible interactions between them. Noting this, we think of computationally exploiting PPI networks to infer protein complexes responsible for individual biological processes.

M.A. Lones et al. (Eds.): IPCAT 2012, LNCS 7223, pp. 250–263, 2012.

Hints for this challenge is found in a different area of technology, web page ranking. The PageRank algorithm used by Google's internet search engine is the most successful web page ranking method [3]. In the PageRank algorithm, each web page is assumed to have an instantaneous value of 'activity'. Activities virtually spread along hyperlinks from pages to pages in the World Wide Web. The amount of activity finally acquired by each page is considered to represent the importance of the page. Higher scores of importance are assigned to pages that are linked from more numerous and more important pages. The web page ranking is hence obtained by sorting web pages in descending order of their activities.

Spreading activation in the PageRank algorithm is mathematically described by a Markov chain in a finite connected graph, which gives a unique steady state. The obtained web page ranking is therefore unique and does not reflect user's personal interest or context. The personalized PageRank (PPR) algorithm is a modified version of the PageRank algorithm and achieves personalized ranking of web pages [3, 4]. In the PPR algorithm, personal interest of a user, namely, what he/she wants to know is expressed by a set of putative web pages, here we call *seed* pages. Activities not only spread along genuine hyperlinks, but 'warp' to the seed pages at a uniform rate as well. Such warp transitions serve as a bias towards the seed pages and make activities tend to localize at and around them. Thus the distribution of activities over the World Wide Web in the steady state reflects user's interest expressed by the seed pages. Web pages with large activities in the steady state are considered to be highly relevant to user's interest.

Recently, an alternative personalized ranking method has been proposed by the present author [5]. Spreading activation in this method is described by continuous-attractor dynamics (CAD) [6]. The CAD was originally developed in computational neuroscience to account for the neural mechanism of graded persistent activity, a type of neuronal activity characterized by its strength that gradually changes in response to transient cue signals [7-9]. Later we have hypothesized that CAD is the mechanism underlying memory retrieval in the brain [6]. We have further pointed out an analogy between memory retrieval with transient cue signals and personalized web page ranking with seed pages [5]. On the basis of this analogy, personalized ranking of documents by the CAD algorithm applied to a citation network has been demonstrated [5]. We have also shown that the performance of personalized document ranking by the CAD algorithm is considerably higher than that by the PPR algorithm [5].

Now we turn to PPI data. Compare a PPI network to the World Wide Web, proteins to web pages and protein-protein interactions to hyperlinks. These comparisons lead to the idea of applying the methods for personalized web page ranking to PPI networks. Indeed, Ivan and Grolmusz have recently examined application of the PPR algorithm to a human PPI network [10]. Taking 'melanoma' as a specific example, they have demonstrated that proteins highly ranked by the PPR algorithm are actually relevant to 'melanoma'.

In this study we propose to infer the relevance of individual proteins to a given biological process (henceforth, referred to as *personalized ranking of proteins* or *personalized protein ranking*) by the CAD algorithm applied to a PPI network. Since the CAD algorithm has exhibited higher performance than the PPR algorithm in personalized ranking of documents, we expect that the CAD algorithm is more advantageous than the PPR algorithm also in personalized ranking of proteins. To confirm this, we carry out a series of comparison experiments.

2   Methods

2.1   Personalized Protein Ranking by Spreading Activation in PPI Networks

Suppose that we wish to know proteins responsible for a certain biological process. Here we propose to infer such proteins by applying spreading activation for personalized ranking to PPI networks.

Personalized ranking by spreading activation is a class of methods originally developed to achieve web page ranking that reflects user's personal interest or context [3, 4]. In these methods, a user initially expresses his/her interest by a set of seed pages. Seed pages are putative web pages that are guessed by the user to be relevant to what he/she wants to know. Reflecting inadequacy of use's knowledge, seed pages might lack some truly relevant pages or include irrelevant ones. Each web page is assumed to have an instantaneous value of activity. These activities virtually spread along hyperlinks in the World Wide Web under the influence of the seed pages (as described later in more detail). This spreading activation finally leads to a steady state where web pages that are truly relevant to user's interest but are absent in the seed pages will have gained certain amounts of activities while those that are irrelevant but are present in the seed pages will have lost their activities [11, 12]. The amount of activity assigned to each web page in the steady state is considered to express the relevance of that page to user's interest. The personalized ranking is hence obtained by sorting web pages in descending order of their activities.

Personalize ranking of proteins can also be accomplished in the same manner. We first express a cellular biological process that we wish to explore by 'seed proteins'. Proteins that have been experimentally identified to be responsible for that biological process, though some of them may be false positive, can be chosen as seed proteins. The relevance of each protein to the biological process is defined by the amount of activity assigned to that protein in the steady state of spreading activation in the PPI network.

2.2   Personalized Ranking as Pattern Restoration

Personalized protein ranking by spreading activation in a PPI network can be compared to a process of pattern restoration. Let ξ_n $(n = 1, \cdots, N)$, with N being the total number of proteins in the PPI network, denote the 'correct' relevance of protein n to a given biological process. Hence we have a 'correct pattern' represented by vector $\vec{\xi} = (\xi_1, \cdots, \xi_N)$. Now let a 'seed pattern' be represented by vector $\vec{\tau} = (\tau_1, \cdots, \tau_N)$: If protein n is a seed protein, $\tau_n = 1$; otherwise $\tau_n = 0$.

In the seed pattern $\vec{\tau}$, some components corresponding to truly relevant proteins might be set $\tau_n = 0$ while those corresponding to irrelevant ones might be $\tau_n = 1$. We can therefore consider $\vec{\tau}$ as a pattern that is generated from a correct pattern $\vec{\xi}$ through degradation due to 'noise' reflecting inadequacy of our knowledge. We model this degradation process as follows. First, the correct pattern $\vec{\xi}$ is digitized to $\vec{d} = (d_1, \cdots, d_N)$: Let θ be the L-th largest component of $\vec{\xi}$; if $\xi_n \geq \theta$, $d_n = 1$; otherwise $d_n = 0$. Protein with $d_n = 1$ are called 'truly relevant proteins'. Then a seed pattern $\vec{\tau}$ is generated from $\vec{d}$ by the following probabilistic rule.

(1a)   If $d_n = 1$, $\tau_n = 1$ with probability $1 - p$ and $\tau_n = 0$ with probability p;

(1b)   if $d_n = 0$, $\tau_n = 0$ with probability $1 - q$ and $\tau_n = 1$ with probability q.

The former describes that some truly relevant proteins might be lost in the seed pattern, while the latter states that some irrelevant ones might be mixed into the seed pattern.

In general, proteins function in groups or complexes in individual biological processes. Each link of a PPI network expresses the tendency that the pair of proteins connected by that link cooperatively function. In this sense, the link structure of a PPI network expresses 'prior knowledge' about such cooperative relations between proteins. On the other hand, a seed pattern $\vec{\tau}$ serves as 'evidence' used as a clue to finding proteins relevant to the biological process. Inference of the relevance of individual proteins is processed by spreading activation in the PPI network, which is to take into account the prior knowledge, under the influence of the seed pattern serving as evidence. Now let the steady state of this spreading activation be represented by $\vec{x}(\infty) = (x_1(\infty), \cdots, x_N(\infty))$. Here, $x_n(\infty) \equiv \lim_{t \to \infty} x_n(t)$ is the activity assigned to protein n in the steady state and is hence considered as the inferred relevance of that protein to the biological process expressed by $\vec{\tau}$. Remind that $\vec{\xi}$ represents the correct relevance. Mapping from $\vec{\tau}$ to $\vec{x}(\infty)$ can therefore be viewed as a 'restoration' process to infer the correct pattern $\vec{\xi}$ from a degraded pattern $\vec{\tau}$.

Once noting the analogy to pattern restoration, we can think of quantitatively evaluating the reliability of personalized protein ranking by measuring the correlation coefficient between the inferred (restored) pattern $\vec{x}(\infty)$ and the correct pattern $\vec{\xi}$:

$$\gamma(\vec{x}(\infty), \vec{\xi}) = \frac{\sum_{n=1}^N (x_n(\infty) - \bar{x}(\infty))(\xi_n - \bar{\xi})}{\sqrt{\left(\sum_{n=1}^N (x_n(\infty) - \bar{x}(\infty))^2\right)\left(\sum_{n=1}^N (\xi_n - \bar{\xi})^2\right)}}, \quad (2)$$

where $\bar{\xi} \equiv \sum_{n=1}^N \xi_n / N$ and $\bar{x}(\infty) \equiv \sum_{n=1}^N x_n(\infty) / N$. The $\gamma(\vec{x}(\infty), \vec{\xi})$ expresses similarity between $\vec{x}(\infty)$ and $\vec{\xi}$. Especially when $\gamma(\vec{x}(\infty), \vec{\xi}) = 1$, the inferred (restored) pattern $\vec{x}(\infty)$ exactly matches the correct pattern $\vec{\xi}$; that is, the restoration is perfect.

2.3   Personalized PageRank (PPR) Algorithm

This study examines personalized protein ranking by using two major methods of spreading activation. The first one is the personalized PageRank (PPR) algorithm [3, 4], defined by the formula

$$x_n(t+1) = \rho I_n(t) + (1-\rho)\vec{\tau} / \sum_{n'=1}^{N} \tau_{n'} \quad (0 \le \rho \le 1), \tag{3}$$

where $x_n(t)$ is the amount of activity assigned to protein n at time t ; $I_n(t) \equiv \sum_{m=1}^{N} T_{nm} x_m(t)$, with $T_{nm} \equiv A_{nm} / \sum_{l=1}^{N} A_{lm}$, is the 'input' to protein n at time t . Here, $\mathbf{A} = (A_{nm})$ is the adjacency matrix of the PPI network: If proteins n and m are directly linked each other, $A_{nm} = A_{mn} = 1$; otherwise $A_{nm} = A_{mn} = 0$. Diagonal elements of $\mathbf{A}$ are vanishing, $A_{nn} = 0$.

    The first term in the R.H.S. of Eq. (3) describes propagation of activities along genuine links, while the second term represents 'warp' of activities to the seed proteins and therefore serves as a bias towards them. Relative weight of contribution of these two is controlled by parameter ρ . After update from $\vec{x}(t)$ to $\vec{x}(t+1)$ by Eq. (3) is repeated many times ($t \to \infty$), $\vec{x}(t)$ converges to a steady state $\vec{x}(\infty)$.

2.4   Continuous-Attractor Dynamics (CAD) Algorithm

The second method of spreading activation for personalized ranking examined in this study is the continuous-attractor dynamics (CAD) algorithm [5, 6]. The CAD was first introduced to account for neural mechanisms of 'graded persistent activity', which is a type of neuronal activity recorded from behaving animals [7, 8] as well as from in vitro slice [9] and is characterized by its strength that gradually changes in response to transient cue signals. This property of graded persistent activity can be modeled by dynamical systems with attractors that continuously depend on the initial state [13]. We have recently proposed that CAD is the mechanism underlying retrieval of short-term memory (expressed by temporal activation of neurons) from long-term memory (stored in synapses) [6].

    Spreading activation in the CAD algorithm is defined by the following rule.

(4a)   If $x_n(t) < I_1$, $x_n(t+1) = I_n(t)/\alpha$;

(4b)   if $I_1 \le x_n(t) \le I_2$, $x_n(t+1) = x_n(t)$;

(4c)   if $I_2 < x_n(t)$, $x_n(t+1) = \alpha I_n(t)$.

Here, $x_n(t)$ and $I_n(t)$ are defined in the same way as in the PPR algorithm, and $I_1 \equiv \alpha I_n(t)$ and $I_2 \equiv I_n(t)/\alpha$ with $0 \le \alpha \le 1$. Update from $\vec{x}(t)$ to $\vec{x}(t+1)$ by this rule is diagrammatically represented by the hysteretic relationship shown in Fig. 1. This hysteretic relationship was originally proposed as the response property of a single cell [9, 14] or an ensemble of cells [15, 16] to produce robust continuous attractors.

In the CAD algorithm, we choose the initial condition as $\vec{x}(0) = \vec{\tau}$. Because of the hysteretic relationship, $\vec{x}(t)$ converges to the steady state $\left(\vec{x}(\infty) \equiv \lim_{t \to \infty} \vec{x}(t)\right)$ that continuously depends on the initial state.

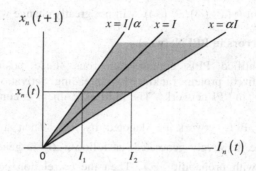

Fig. 1. Diagrammatic representation of the hysteretic input/output relationship defined by the rule (4a-c) in the text

2.5 Comparison Experiment

To determine which algorithm, the PPR or CAD algorithm, is more efficient for personalized ranking of proteins, a series of comparison experiments are conducted.

I Performance of Personalized Protein Ranking

In the first experiment, the performance of personalized protein ranking is compared between the PPR algorithm and the CAD algorithm. A correct pattern $\vec{\xi}$ is synthesized on the PPI network and a seed pattern $\vec{\tau}$ is generated from $\vec{\xi}$ according to the degradation processes (1). Then the correlation coefficient between $\vec{x}(\infty)$ inferred either by the PPR or CAD algorithm and $\vec{\xi}$ is examined at each value of the unique parameter in either algorithm (ρ for the PPR algorithm and α for the CAD algorithm, both ranging from zero to one).

II Robustness against Change in the Seed Pattern

Choice of seed proteins to express a biological process that we wish to explore might be associated with ambiguity reflecting inadequacy of our knowledge. Nevertheless, the result of personalize protein ranking should not be greatly affected by such ambiguity. Hence the second experiment is conducted to examine the robustness of personalize protein ranking against changes in the seed pattern.

Now let $\vec{\tau}^{(0)}$ be a seed pattern generated from a correct pattern $\vec{\xi}$ by the degradation processes (1). From $\vec{\tau}^{(0)}$, a series of gradually changing patterns $\vec{\tau}^{(i)}$ $(i = 1, 2, \cdots)$ are further generated, as follows: If $\tau_n^{(i-1)} = 1$, $\tau_n^{(i)} = 0$ with probability p_S and $\tau_n^{(i)} = 1$ with probability $1 - p_S$; otherwise (that is, if $\tau_n^{(i-1)} = 0$), $\tau_n^{(i)} = 0$. Then the correlation coefficient between $\vec{x}^{(0)}(\infty)$, the pattern inferred from $\vec{\tau}^{(0)}$, and $\vec{x}^{(i)}(\infty)$, that inferred from $\vec{\tau}^{(i)}$, is examined as a function of i. If personalized protein ranking is robust against changes in the seed pattern, the correlation coefficient $\gamma(\vec{x}^{(0)}(\infty), \vec{x}^{(i)}(\infty))$ will not greatly change with i.

III   Stability to Errors in PPI Networks

Experimentally identified PPIs contain numerous 'false positives' and 'false negatives'. Personalized protein ranking by spreading activation is expected to tolerate such 'errors' in PPI networks. The third experiment is conducted to confirm this expectation.

Let the original PPI network be denoted by G_0. Then a series of graphs G_k $(k = 1, 2, \cdots)$ are iteratively generated, as follows: G_k is generated by deleting each link of G_{k-1} with probability p_G. Then the correlation coefficient between $\vec{x}^{[0]}(\infty)$, the pattern inferred from a seed pattern $\vec{\tau}$ in G_0, and $\vec{x}^{[k]}(\infty)$, that inferred from the same seed pattern in G_k, is examined as a function of k. If personalized protein ranking by spreading activation is stable to errors in PPI networks, the correlation coefficient $\gamma(\vec{x}^{[0]}(\infty), \vec{x}^{[k]}(\infty))$ will not greatly change with k.

2.6   Yeast Protein-Protein Interaction Network

The above experiments are done by using a yeast protein-protein interaction (YPPI) network constructed from published data downloaded from [17]. The YPPI network comprises $N = 3728$ proteins as nodes and 4759 protein-protein interactions as undirected links.

2.7   Parameter Values

Unless otherwise stated the following parameter values are used in the present study: $L = 100$; $p_S = 0.2$; $p_G = 0.1$; $p = 0.5$; $q = 0.2(1 - p)L/N$.

3   Results

3.1   Personalized Protein Ranking by Either the PPR or CAD Algorithm

We propose to infer the relevance of proteins to a given cellular biological process ('personalized ranking of proteins' or 'personalized protein ranking') using either the PPR or CAD algorithm applied to PPI networks. To determine which algorithm is more efficient for personalized ranking of proteins, we carried out three comparison experiments using YPPI data.

3.2   Comparing the Performance of Personalized Protein Ranking between the PPR Algorithm and the CAD Algorithm

The first experiment was conducted to compare the performance of personalized protein ranking between the PPR algorithm and the CAD algorithm. The correlation coefficient between a correct pattern $\vec{\xi}$ and the pattern $\vec{x}(\infty)$ inferred by either algorithm was calculated. Then the correlation coefficient averaged over 100 $\vec{\xi}$'s was plotted as a function of the value of the unique parameter in each algorithm (ρ for the PPR algorithm and α for the CAD algorithm). The average correlation coefficient for the PPR algorithm peaks at $\rho = 0.8$, while that for the CAD algorithm exceeds this peak value for a wide range of α ($0.4 \leq \alpha \leq 0.7$) (Fig. 2). These results demonstrate that the performance of personalized protein ranking by the CAD algorithm is considerably higher than that by the PPR algorithm.

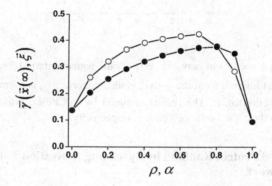

Fig. 2. The correlation coefficient between the inferred pattern $\vec{x}(\infty)$ and the correct pattern $\vec{\xi}$ is averaged over 100 ξ 's and plotted as a function of either ρ (filled circle) or α (open circle)

3.3   Comparing the Robustness against Changes in the Seed Pattern

Choice of seed proteins used as a clue to personalized protein ranking for a given biological process might be associated with a certain degree of ambiguity. Nevertheless, the result of personalized protein ranking should not be too sensitive to exact choice of seed proteins. Hence we conducted the second experiment to compare the robustness against changes in the seed pattern between the two algorithms.

A series of gradually changing patterns $\vec{\tau}^{(0)}, \vec{\tau}^{(1)}, \vec{\tau}^{(2)}, \cdots$ were generated as described in **Methods**. We examined how the correlation coefficient between the pattern inferred for $\vec{\tau}^{(0)}$, denoted by $\vec{x}^{(0)}(\infty)$, and that for $\vec{\tau}^{(i)}$, denoted by $\vec{x}^{(i)}$ $(i = 1, 2, \cdots)$, changes with i. As i increases, the correlation coefficient

$\gamma\left(\vec{x}^{(0)}(\infty), \vec{x}^{(i)}(\infty)\right)$ for the PPR algorithm more progressively decays than that for the CAD algorithm (Fig. 3). These results demonstrate that personalized protein ranking by the CAD algorithm is more robust against ambiguity associated with the choice of seed proteins than that by the PPR algorithm.

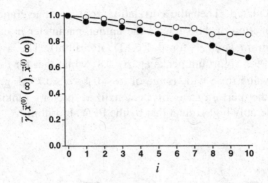

Fig. 3. The correlation coefficient between $\vec{x}^{(0)}(\infty)$, the pattern inferred from the seed pattern $\vec{\tau}^{(0)}$, and $\vec{x}^{(i)}(\infty)$, that inferred from the i-th degraded pattern $\vec{\tau}^{(i)}$, is averaged over 100 ξ's and plotted as a function of i. The results obtained by the PPR algorithm and the CAD algorithm are indicated by filled and open circles, respectively.

3.4   Personalized Protein Ranking by Spreading Activation Tolerates Errors in PPI Networks

PPI data contain numerous false positives and negatives. Personalized protein ranking by spreading activation is expected to tolerate such errors in PPI networks. We conducted the third experiment to confirm this expectation.

A series of graphs $G_1, G_2, \cdots$, the structures of which are gradually changing, were generated from the original YPPI network (denoted by G_0) as described in **Methods.** The correlation coefficient between $\vec{x}^{[0]}(\infty)$, the pattern inferred for a seed pattern $\vec{\tau}$ in G_0, and $\vec{x}^{[k]}(\infty)$, that inferred for the same seed pattern in G_k, was examined as a function of k. As k increase, the correlation coefficient for either the PPR or CAD algorithm only slightly decays, and there is little difference in the decay rate between the two algorithms (Fig. 4). For instance, more than half of the links of the original YPPI network are deleted at $k=8$ $(1-(1-p_G)^8 \sim 0.57$ for $p_G=0.2$, see **Methods**). Nevertheless, the correlation coefficient for either algorithm is still considerably high (~0.8). These results suggest that personalized protein ranking by spreading activation can generally tolerate errors in PPI networks to some extent.

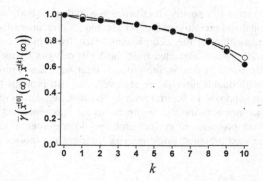

Fig. 4. The correlation coefficient between $\vec{x}^{[0]}(\infty)$, the pattern inferred in graph G_0, and $\vec{x}^{[k]}(\infty)$, that inferred in the k-th degraded graph G_k is averaged over 100 ξ's and plotted as a function of k. The results obtained by the PPR algorithm and the CAD algorithm are indicated by filled and open circles, respectively.

3.5   Personalized Protein Ranking for a Specific Example

Finally we demonstrate personalized protein ranking for a specific example, 'positive regulation of spindle pole body separation [18]', by either the CAD or PPR algorithm. This demonstration highlights the difference in the way of inference between the two algorithms and reveals their characteristics.

The proteins annotated to 'positive regulation of spindle pole body separation' in the public database [18] were supposed as 'correct' proteins, which are listed in Table 1a. Seed proteins, listed in Table 1b, were generated from the correct proteins as follows: Three proteins (marked by double asterisk in Table 1a) were removed from the correct protein list, and then two randomly chosen proteins (marked by single asterisk in Table 1b), which are not annotated to 'positive regulation of spindle pole body separation', were added as 'noise'. We examined whether personalized protein ranking by either of the two algorithms could restore the correct proteins missing in the seed pattern and filter out the noise proteins.

The top 20 of personalized protein ranking by the CAD algorithm is shown in Table 1c, where the correct proteins are shaded by grey. One of the two proteins missing in the seed pattern was restored (indicated by double asterisk in Table 1c), while the noise proteins were completely eliminated from the top 20. In contrast, the top 20 of personalized protein ranking by the PPR algorithm (Table 1d) could not retrieve any of the missing proteins. Furthermore, the proteins added as noise in the seed pattern were still present and highly ranked in the top 20 (indicated by single asterisk in Table 1d).

At a glance of Table 1d, one might think that the PPR algorithm is favorable, as the correct proteins in the seed pattern are more highly ranked by the PPR algorithm than by the CAD algorithm. Note however that the noise proteins are also highly ranked by the PPR algorithm. These just reflect the fact that personalized protein ranking by the PPR algorithm is strongly influenced by the choice of seed proteins, as generally demonstrated in Section 3.3.

Table 1. (a) Correct proteins for 'positive regulation of spindle pole body separation'. (b) Seed proteins used for personalized protein ranking. (c) The top 20 proteins extracted from the YPPI network by the CAD algorithm for the seed proteins in (b), listed in descending order of their activities. (d) The top 20 proteins extracted from the YPPI network by the PPR algorithm for the seed proteins in (b), listed in descending order of their activities. From the list of correct proteins in (a), those with double asterisk are removed, and two proteins (indicated by single asterisk in (b)) that are supposed to be irrelevant to 'positive regulation of spindle pole body separation' are added as 'noise' to the list; seed proteins are thus generated and listed in (b). In (c) and (d), grey shaded proteins are correct proteins; double asterisk indicates the correct proteins missing in the seed proteins; single asterisk indicates the 'noise' proteins mingled in the seed protein.

(a)	
CDC28	**
CDC5	**
CLB1	
CLB2	
CLB3	
CLB4	**
CLB5	

(b)	
CLB1	
CLB2	
CLB3	
CLB5	
YML126C	*
KRE5	*

(b)		
protein	activity	
JSN1	0.140652	
CLB2	0.103928	
CKS1	0.079191	
SRB4	0.063575	
NIP29	0.04396	
LYS14	0.031747	
CLB1	0.028906	
YOR264W	0.026397	
RPS11A	0.026397	
SCO2	0.026397	
CLB3	0.014219	
SHO1	0.013862	
YOR285W	0.007358	
YAR027W	0.006122	
YDR078C	0.006056	
DUT1	0.005987	
CLB4	0.005982	
CDC28	0.005656	**
GCS1	0.004918	
NUP53	0.004892	

(d)		
protein	activity	
JSN1	0.099257	
CLB2	0.050036	
SRB4	0.04553	
CLB1	0.038755	
CLB3	0.037346	
CKS1	0.033887	
KRE5	0.033705	*
YML126C	0.033612	*
CLB5	0.033612	
NIP29	0.027106	
LYS14	0.017877	
YOR264W	0.017163	
SHO1	0.011183	
RPS11A	0.010688	
SCO2	0.010007	
ATP14	0.006016	
YOR285W	0.005959	
BZZ1	0.005427	
YDR078C	0.004825	
YAR027W	0.004743	

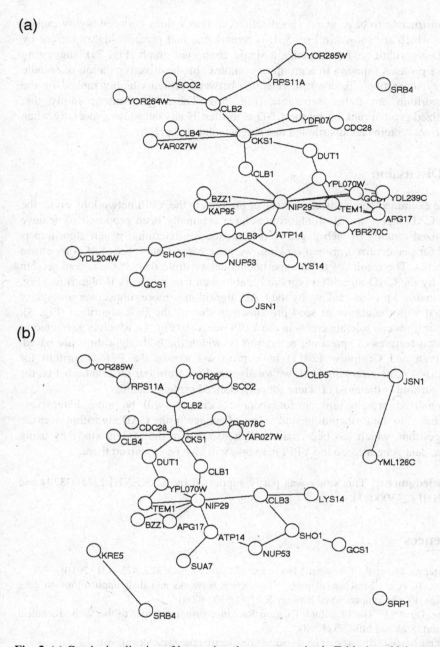

Fig. 5. (a) Graph visualization of interactions between proteins in Table 1c, which are extracted by the CAD algorithm. (b) Graph visualization of interactions between proteins in Table 1d, extracted by the PPR algorithm.

It is instructive to have graph visualizations of interactions between highly ranked proteins, which are shown in Fig. 5. It is remarkable that proteins highly ranked by the CAD algorithm tend to organize a single connected graph (Fig. 5a), suggesting that these proteins function in a group or complex for 'positive regulation of spindle pole body separation'. In contrast, relations between proteins highly ranked by the PPR algorithm are rather separable (Fig. 5b). These observations imply that personalized protein ranking by the CAD algorithm is advantageous especially when most correct proteins are distributed in a single connected graph.

4   Discussion

We have examined personalized ranking of proteins in the YPPI network by either the PPR or CAD algorithm. Both algorithms have originally been proposed to achieve personalized ranking of web pages or documents. To determine which algorithm is the best for personalized protein ranking, we have carried out a series of comparison experiments. The results of these experiments demonstrate that: Personalized protein ranking by the CAD algorithm is more reliable than that by the PPR algorithm (Fig. 2); personalized protein ranking by the CAD algorithm is more robust over ambiguity associated with the choice of seed proteins than that by the PPR algorithm (Fig. 3); both algorithms can tolerate errors in the YPPI network (Fig. 4), which is attributed to the general features of spreading activation on which the both algorithms are based. While Ivan and Grolmusz (2011) have proposed to use the PPR algorithm for personalized protein ranking [10], we totally conclude from our observations that the CAD algorithm is the most efficient for personalized protein ranking.

Personalized protein ranking for specific examples will be more extensively demonstrated in the forthcoming study. The advantage of the CAD algorithm over the PPR algorithm, which has been mainly demonstrated in the present study by using synthetic data generated on the YPPI network, will also be examined there.

Acknowledgments. This study was partly supported by KAKENHI (23500379) and KAKENHI (23300061).

References

1. Costanzo, M., et al.: The genetic landscape of a cell. Science 327, 425–431 (2010)
2. Hase, T., et al.: Structure of protein interaction networks and their implications on drug design. PLoS Computational Biology 5, e1000550 (2009)
3. Page, L., et al.: The PageRank Citation Ranking: Bringing Order to the Web. Technical Report, Stanford InfoLab (1998),
   http://www-db.stanford.edu/~backrub/pageranksub.ps
4. Haveliwala, T.: Topic-Sensitive PageRank: A Context-Sensitive Ranking Algorithm for Web Search. IEEE Trans. Knowledge Data Eng. 15, 784–796 (2003)
5. Okamoto, H.: Topic-Dependent Document Ranking: Citation Network Analysis by Analogy to Memory Retrieval in the Brain. In: Honkela, T. (ed.) ICANN 2011, Part I. LNCS, vol. 6791, pp. 371–378. Springer, Heidelberg (2011)

6. Tsuboshita, Y., Okamoto, H.: Graded information extraction by neural-network dynamics with multihysteretic neurons. Neural Netw. 22, 922–930 (2009)
7. Romo, R., et al.: Somatosensory discrimination based on cortical microstimulation. Nature 399, 470–473 (1999)
8. Aksay, E., et al.: In vivo intracellular recording and perturbation of persistent activity in a neural integrator. Nat. Neurosci. 4, 184–193 (2001)
9. Egorov, A.V., et al.: Graded persistent activity in entorhinal cortex neurons. Nature 420, 173–178 (2002)
10. Ivan, G., Grolmusz, V.: When the Web meets the cell: using personalized PageRank for analyzing protein interaction networks. Bioinformatics 27, 405–407 (2011)
11. Collins, A.M., Loftus, E.F.: Spreading-Activation Theory of Semantic Processing. Psychol. Rev. 82, 407–428 (1975)
12. Anderson, J.R., Pirolli, P.L.: Spread of activation. J. Exp. Psychol. 10, 791–798 (1984)
13. Seung, H.S., et al.: Stability of the memory of eye position in a recurrent network of conductance-based model neurons. Neuron 26, 259–271 (2000)
14. Goldman, M.S., et al.: Robust persistent neural activity in a model integrator with multiple hysteretic dendrites per neuron. Cereb. Cortex 13, 1185–1195 (2003)
15. Koulakov, A.A., et al.: Model for a robust neural integrator. Nat. Neurosci. 5, 775–782 (2002)
16. Okamoto, H., et al.: Temporal integration by stochastic recurrent network dynamics with bimodal neurons. J. Neurophysiol. 97, 3859–3867 (2007)
17. Yeast Interacting Proteins Database, Copyright© 2009 Takashi Ito (the University of Tokyo) licensed under CC Attribution-Share Alike 2.1 Japan,
    http://dbarchive.biosciencedbc.jp/en/yeast-y2h/desc.html
18. Saccharomyces Genome Database,
    http://www.yeastgenome.org/cgi-bin/GO/goTerm.pl?goid=10696

A Parallel Algorithm for Multiple Biological Sequence Alignment

Irma R. Andalon-Garcia, Arturo Chavoya*, and M.E. Meda-Campaña

Department of Information Systems, Universidad de Guadalajara,
Periferico Norte 799-L308, Zapopan, Jal., Mexico 45100
{agi10073,achavoya,emeda}@cucea.udg.mx
http://www.cucea.udg.mx

Abstract. The search of a multiple sequence alignment (MSA) is a well-known problem in bioinformatics that consists in finding a sequence alignment of three or more biological sequences. In this paper, we propose a parallel iterative algorithm for the global alignment of multiple biological sequences. In this algorithm, a number of processes work independently at the same time searching for the best MSA of a set of sequences. It uses a Longest Common Subsequence (LCS) technique in order to generate a first MSA. An iterative process improves the MSA by applying a number of operators that have been implemented to produce more accurate alignments. Simulations were made using sequences from the UniProKB protein database. A preliminary performance analysis and comparison with several common methods for MSA shows promising results. The implementation was developed on a cluster platform through the use of the standard Message Passing Interface (MPI) library.

Keywords: Multiple sequence alignment, Parallel processing, Longest common subsequence, Dynamic programming.

1 Introduction

Bioinformatics is defined by Luscombe et al. as a discipline that applies computational techniques to the analysis of biological information on a large scale [16]. The success of bioinformatics is partly due to the fact that genomes of many species —including our own— have been sequenced, as well as to the emergence of new and faster technologies to gather data from large and complex biological sequences. The existence of numerous public biological databases that can be accessed through the World Wide Web has also contributed to this success. Considering that the number of genetic sequences is growing in size exponentially as time passes, improving algorithms used in bioinformatics is nowadays a necessity.

A basic subarea of bioinformatics is biological sequence alignment and analysis [18], which compares and finds similarities in nucleic acid (DNA and RNA) and amino acid sequences in order to classify and identify homologies or patterns

---

* Corresponding author.

M.A. Lones et al. (Eds.): IPCAT 2012, LNCS 7223, pp. 264–276, 2012.

among genes and proteins, as well as to predict biological functions and structures. The alignment of three or more sequences is known as Multiple Sequence Alignment (MSA).

The aim of this paper is to propose a parallel algorithm for the global alignment of multiple biological sequences (PaMSA). A well-known Longest Common Subsequence (LCS) algorithm is used to seed a first alignment of the sequences. Sequences are stored and handled in memory as strings. After this step, an iterative process improves the MSA by applying a number of operators that have been implemented. The proposed operators are focused on the detection of gaps and identical or similar residues that are not totally aligned. These string operators perform an exhaustive search along the total length of all sequences with the aim of finding an opportunity to improve the alignment. Several processes work independently at the same time searching for the best MSA of a set of sequences. There exists a process that acts as a coordinator, whereas the rest of the processes are considered slave processes. Each slave process sends its MSA score to the coordinator, which selects the best MSA (Fig. 1). In order to test the algorithm, simulations were made using similar groups of sequences from the UniProKB protein database located at http://www.uniprot.org/. A preliminary performance analysis and comparison with several common methods rendered encouraging results.

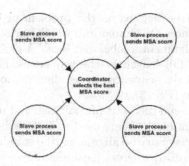

Fig. 1. Communication topology used among processes

1.1 Biological Sequences

A biological sequence is a single, continuous molecule of nucleic acid or protein. Sequences are represented as a succession of letters and customarily they are written with no gaps (Fig. 2). The letters used to symbolize the different bases of a DNA strand are A, C, G, and T, representing the four nucleotide bases —adenine, cytosine, guanine and thymine, respectively—. The nucleic acid sequence of a gene can be translated into a protein sequence (chain of amino acids). Each of the twenty different amino acids is represented by a letter from the alphabet {A,C,D,E,F,G,H,I,K,L,M,N,P,Q,R,S,T,V,W,Y}. Protein sequences are the fundamental determinants of biological structure and function.

At present, biological databases have become an important tool in life sciences. They contain protein and gene sequences, published literature, and other important data used to understand biological phenomena. Bioinformatics uses this kind of data for medical and biological research. An important resource to locate biological data-bases is the Nucleic Acids Research (NAR) Database Issue and the online Molecular Biology Database Collection. The current 18th annual Database Issue (2011) provides 1330 selected databases covering several aspects of biological research [6].

...AGACTTAGACTCAGGA...

DNA sequence

...EDKCLSTENSEDKCES...

Protein sequence

Fig. 2. Examples of biological sequences

1.2 Sequence Alignment

Sequence alignment can be defined as the problem of comparing and finding which parts of the sequences are similar and which parts are different. There are two types of sequence alignment: global and local. A global alignment optimizes the alignment over the full length of the sequences, i.e. it matches as many characters as possible in the sequences to be aligned. Local alignment looks for matches in a local region (Fig. 3).

There exist computational methods for pairwise and multiple sequence alignments. Pairwise alignment methods are used to find the best matching alignment of two sequences. For global pairwise alignment, the standard Needleman-Wunsch [20] algorithm can be applied, whereas local alignments can be obtained via the Smith-Waterman [24] algorithm. The set of sequences to be aligned are assumed to have an evolutionary relationship. Homologous sequences are sequences that share a common ancestor and usually also share common functions.

1.3 Related Work

The strategies used to align multiple sequences are more complex and sophisticated than the techniques used for pairwise alignment. There are several algorithms that can be used in sequence alignment [9]. A number of strategies have been applied to improve MSAs, such as progressive alignments methods [25], iterative methods [7], a simulated annealing method [10], genetic algorithms [21], greedy algorithms [29] and Markov chain processes [4].

Currently there exist popular computer programs used to find multiple sequence alignments, such as MUSCLE [5], T-Coffee [22], and ClustalW [25].

A G A A T - C T G C A G   A G A A T T C A G C A G

A - A G T A C T G T A G   - - - A T G C A - - - -

Global alignment   Local alignment

Fig. 3. Global alignment vs. local alignment

In addition, some computer applications have been designed to explore the available sequence databases. One example is FASTA [14], a well-known and widely used DNA and protein sequence alignment software package. The Basic Local Alignment Search Tool (BLAST) [1], a set of programs for comparing biological sequences, is another example.

On the other hand, High Performance Computing (HPC) is widely used at present in the development of applications that require solving complex problems, that need long processing times, or that handle a large amount of data. MSA algorithms are a typical example of this class of applications. Consequently, research efforts have been focused on developing faster algorithms for finding MSAs exploiting the multiprocessor capabilities of modern computers [2][11][13]. Recent efforts for improving accuracy of multiple sequence alignment have also been done by combining various methods or tools [28][3][27][17][15][12][23].

2   Method

Given a set of n biological sequences $A_1, A_2, \ldots, A_n$, where $A_i = a_{i1}, a_{i2}, \ldots, a_{im}$ and m is the length of A_i, for $i = 1 \ldots n$, the proposed approach uses an iterative parallel method for finding the MSA of the n sequences in the set. Let np be the number of processes used by the algorithm. The number of possible different MSA solutions is equal to np, for $np = 1, 2, \ldots, n$. For example, if $np = 2$, there will be 2 independent processes searching for the best MSA. As the number of processes increases, the number of potential solutions rises.

At first, each process —including the process acting as coordinator— reads the sequences from a file and stores the sequences into a memory array. Sequences in the file must be in the FASTA format, which is described in [19]. The main steps that are performed for each parallel process are described next. Each process has a consecutive integer ID, which acts as an identification number for the process.

2.1   Applying the Longest Common Subsequence Algorithm

At step 1, a well-known LCS technique for two sequences [26] is used in order to generate the first alignment of the sequences in each parallel process.

Given a sequence $A = a_1 a_2 \ldots a_m$, a subsequence of A is a sequence $C = c_1 c_2 \ldots c_p$, defined by $c_k = a_{n_k}$ where m is the length of sequence A, as defined previously, $n_1 < n_2 < n_p$, p is the number of selected items from sequence A,

$1 \leq k \leq p$ and $p \leq m$, i.e. C can be obtained by deleting $m - p$ (not necessarily contiguous) symbols from A without changing its order.

The sequence $C = c_1 c_2 \ldots c_p$ is a common subsequence of the sequences $A = a_1 a_2 \ldots a_m$ and $B = b_1 b_2 \ldots b_n$ if C is a subsequence of both sequences (A and B). For instance, a common subsequence of "ATC" and "TAC" is "AC", but also is "TC". It is not uncommon to have more than one common subsequence.

The longest common subsequence of A and B, is a longest sequence C, which is both a subsequence of A and B. In general, the LCS problem consists of finding the maximal length subsequence —i.e. there exists no other subsequence that has greater length— that is a common subsequence of all the sequences in the set. It is possible to have more than one LCS.

Let A and B be the above defined sequences of length m and n, respectively. The algorithm that we use to obtain the LCS of two sequences uses dynamic programming and requires calculating the LCS table (LCST) following these rules:

$$LCST(i,j) = \begin{cases} 0 & \text{if } j = 0 \quad \text{or} \quad i = 0 \\ LCST(i-1, j-1) + 1 & \text{if } i > 0, j > 0 \text{ and } a_i = b_j, \\ max(LCST(i, j-1), LCST(i-1, j)) & \text{if } i > 0, j > 0 \text{ and } a_i \neq b_j \end{cases}$$

where $i = 1, \ldots, m$ and $j = 1, \ldots, n$, m is the number of rows in LCST, n is the number of columns in LCST, and the cell $LCST(i,j)$ is the element in the LCS table at row i, column j. The LCS table stores numbers which correspond to the actual length of the LCS. At the end the last cell in the LCS table contains the length of the LCS. The longest common subsequence can be found by backtracking from the cell at $LCST(m, n)$. Each time a match is found, it is appended to the longest common subsequence. In other case, a movement is made to the cell with $max(LCST(i-1, j), LCST(i, j-1), LCST(i-1, j-1))$ in order to find the next match.

For example, let $A=$"AT" and $B=$"ACT". After application of the previous rules, the completed LCS table (LCST) generated is shown in Fig. 4, and the subsequence "AT" is the LCS of the sequences A and B. In general, there may be several such paths because the longest common subsequence is not necessarily unique. The cell $LCST(m, n)$ contains the length of the LCS of A and B.

It should be mentioned that we do not use this algorithm to obtain the longest common subsequence; instead, results in the LCS table are used to generate a first MSA. The goal is to align identical residues in the sequences. The procedure is as follows: each process applies the LCS algorithm to all the possible pairs of sequences in the set that result from the combination of sequence n (for process $ID = n$) with the rest of the sequences. For example, the process with $ID = 1$ applies the algorithm to the following pair of sequences: $LCS(A_1, A_2), LCS(A_1, A_3), \ldots, LCS(A_1, A_n)$. With the result of each calculation, a pairwise alignment is generated, which detects all the identical residues of the sequences. Next, a matrix is created containing the results of all calculations. Finally, gaps are inserted if necessary at the end of sequences in order

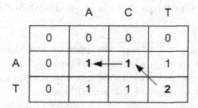

Fig. 4. The resulting LCS table for the AT and ACT sequences

to have all the sequences with the same length. As a result, we obtain an initial MSA. The process with $ID = 2$ will apply the algorithm to the following pair of sequences: $LCS(A_2, A_1), LCS(A_2, A_3), \ldots, LCS(A_2, A_n)$, and the same steps described previously for the process with $ID = 1$ take place for this process, and the rest of the processes. Even though all processes apply the same algorithm to the same set of sequences, the resulting MSAs are possibly different, because the calculations are based on a different sequence and there exists the possibility of having more than one LCS for the same two sequences (Fig. 5). This procedure allows various potential solutions running in distinct processes.

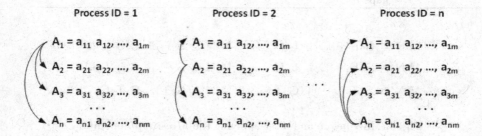

Fig. 5. Dynamics of applying the LCS algorithm

2.2 Parallel Iterative Algorithm

After the first MSA is generated in each process, the alignment is improved by applying a number of operators in an iterative manner. Currently, there exist ten operators defined, which are shown in Table 1. These proposed operators perform an exhaustive search along the total length of all sequences with the aim of finding an opportunity to improve the alignment. The search is focused on the detection of gaps and identical or similar residues that are not totally aligned. At each iteration, operators are applied when necessary. For example, Fig. 6 shows the MSA before applying the operator named $moveGapL$ ("move gap to the left") to the last sequence in the figure, and also illustrates the MSA after the operation.

Table 1. Operators defined in the PaMSA algorithm

Operator	Objective	Scope
moveGapR	Moves a gap in 2nd seq. one space to the right to align identical residues	Two consecutive sequences
moveGapRA	Moves a gap in 2nd seq. one space to the right to align identical residues	First sequence with all others
moveGapRN	Moves a gap in 1st seq. one space to the right to align identical residues	Two consecutives sequences
moveGapRF	Moves a gap in 1st seq. one space to the right to align identical residues	First sequence with all others
moveGapDRN	Moves a gap in 1st. seq.one space to the right to align similar residues	Two consecutive sequences
moveGap2R	Moves a gap in 1st. seq. two spaces to the right to align similar residues	Two consecutive sequences
moveGapS2R	Moves a gap in 2nd. seq. two spaces to the right to align similar residues	Two consecutive sequences
moveGapL	Moves a gap in 2nd. seq. one space to the left to align identical residues	First sequence with all others
shiftGapR	Moves a gap in 2nd. seq. one space to the leftt to send gaps at the end	First sequence with all others
removeEndGaps	Removes the last column of the alignment if all the elements are gaps	All the sequences

```
M N L S F F C L L M          M N L S F F C L L M
M N A S F - C L L M          M N A S F - C L L M
M N A S T C - T G T          M N A S T - C T G T
```

Fig. 6. MSA before and after applying the *MoveGapL* operator

A fundamental property to evaluate the performance of alignment methods is the estimation of the MSA accuracy. MSA accuracy plays a central role in several areas of biology. For example, alignment accuracy is important to determine more accurate species trees —i.e. alignment accuracy impacts directly on phylogenetic inference—. Therefore, there is a continuos effort to obtain more accurate MSAs. In order to evaluate the accuracy of the MSA obtained, the sum-of-pairs (SP) score method has been implemented. The sum-of-pairs score is an accurate metric for an MSA, based on the number of correctly aligned residue pairs, where the score of all pairs of sequences in the multiple alignment is added to the overall score. The sum-of-pairs score for the MSA is calculated as

$$SP(MSA) = \sum_{i=1}^{r} \sum_{k<l} s(m_i^k, m_i^l),$$ (1)

where r is the length of the MSA, $s(m_i^k, m_i^l)$ is the score obtained by comparing the kth row in the ith column of the MSA, with the lth row in the same i^{th} column of the MSA. Each process sends the scoring of its MSA to the coordinator, which receives the scores, selects the process with the best score, and determines the resulting MSA.

Another measure of the accuracy of an MSA is the identity percentage. In our implementation, the identity percentage obtained among the sequences is calculated as

$$ID(MSA) = \left(\left(\sum_{j=1}^{r} \sum_{i=1}^{n} x_i^j \right) * 100 \right) /r,$$ (2)

where r is the length of the MSA, n is the number of sequences aligned, and $\sum_{i=1}^{n} x_i^j$ is counted only if all the x_i residues at the jth column of the MSA are identical.

In addition, we use a substitution matrix in order to obtain better alignments. A substitution matrix is used to describe how fast a character in a sequence changes into another character over time. The BLOcks of Amino Acid SUbstitution Matrix or BLOSUM [8] is commonly used in sequence alignment. The BLOSUM62 matrix was used in our simulations. The algorithm ends if a pre-defined number of iterations is reached, or if the sum-of-pairs score of the MSA has not shown improvement in two consecutive iterations.

2.3 Implementation

The implementation was developed in a cluster provided by Intel Corporation, which contains 10 nodes, each node with 2 Intel Xeon 5670 6-core 2.93 GHz CPUs, 24-GB of 1066 MHz DR3 RAM, and two 274 GB 15K RPM hard drives. The operating system used was Red Hat Enterprise Linux 5 Update 4 with Perceus 1.5 Clustering Software and Server 5.3 running Intel MPI 3.2.

The Message Passing Interface (MPI) library —a standard specification that allows portable message communication in programs written in the Fortran, Ada and C programming languages— was used in our implementation. Portability is one of the main advantages of MPI, as it has been implemented in almost all of the distributed memory architectures. MPI defines the syntax and semantics of a set of functions in a library designed to exploit the existence of multiple processors, and it provides the synchronization and communication needed among processes.

Synchronous communication operations were used in this work to handle communication and synchronization among tasks. When a synchronous operation is invoked, a process sends a message and then waits for a response before proceeding with the process flow. By using synchronous operations, we can be confident that the parallelized algorithm will work correctly. Finally, object-oriented and structured programming paradigms were applied using C++ as the programming language.

3   Simulation and Results

In order to test our parallel implementation of the global alignment algorithm for multiple biological sequences, a number of groups of similar sequences was used. The sequences were obtained from http://www.uniprot.org/ of the *UniProKB* protein database at Sequences Clusters (*UniRef*). At this site, sequences are classified in groups —named clusters— according to their identity percentage, so similar sequences can be obtained through a database query. Sequences that belong to a specific cluster are named cluster members. With the aim of comparing the performance of PaMSA with other MSA methods, the cluster named *UniRef*100_A0A132 at http://www.uniprot.org/uniref/UniRef90_A0A132 was used. This dataset has 23 members with a length of approximately 53 residues. Additionally, for the PaMSA simulations, the cluster named *UniRef*100_A0A1J8 at http://www.uniprot.org/uniref/UniRef90_A0A1J8 with sequence length of approximately 300 residues was used.

Table 2 shows the preliminary execution time results obtained in simulations, which are the average of 5 runs. As can be seen, as the number and length of the sequences increase, execution time rises. However, this increase in response time is not excessive, suggesting that this algorithm could be useful in the MSA analysis of a large number of sequences.

Figure 7 shows the execution times results achieved by PaMSA and various common methods for MSA (Muscle online v3.8.31, ClustalW online v2.0.12, Dialign online v2.2.1 and Muscle v.3.8.31 for Windows). As can be seen, PaMSA obtained the best response times in all the cases tested. The delay in response time of the online MSA versions may be due to the communication protocols used when sending and receiving processes. Nonetheless, PaMSA was superior even when compared to the Windows version of Muscle.

On the other hand, Table 3 shows the preliminary accuracy results obtained in our simulations. As mentioned above, we used clusters of sequences with a known identity of approximately 90%. Thus, the results obtained are not optimal, but seem to be promising with respect to accuracy.

It should be mentioned that if fewer sequences are aligned, the identity percentage can be above 90% if the sequences compared are very similar; the more sequences are aligned, the less identity percentage is expected due to increased sequence divergency.

Figure 8 shows the accuracy results achieved by PaMSA and the various common methods for MSA tested. The PaMSA algorithm is less accurate than the other methods. However, the accuracy results obtained by the PaMSA algorithm were almost equal to the results of the other methods tested.

Table 2. Execution times (seconds) of the PaMSA algorithm

Number of sequences	Length of the sequences (approx.)		
	100 residues	150 residues	300 residues
3	0.005	0.007	0.018
5	0.017	0.021	0.043
10	0.045	0.120	0.278
20	0.172	0.273	0.658
50	1.336	1.367	2.899

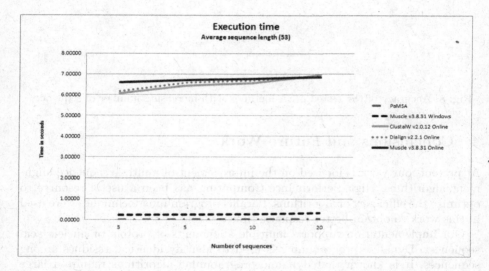

Fig. 7. Execution times of the tested MSA methods with increasing number of sequences

Table 3. Accuracy of the PaMSA algorithm on protein alignments

Number of sequences	Length of the sequences (approx.)		
	100 residues	150 residues	300 residues
3	87.7%	91.6%	97.3%
5	84.2%	89.7%	95.6%
10	80.7%	74.8%	92.9%
20	68.4%	74.8%	89.2%
50	64.9%	74.8%	89.2%

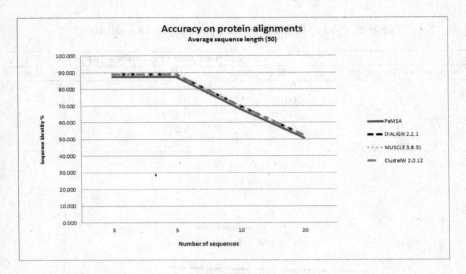

Fig. 8. Accuracy of the tested MSA methods with increasing number of sequences

4   Conclusions and Future Work

At present, our work is focused on the improvement of multiple sequence alignment algorithms. High Performance Computing has been a useful resource to optimize the efficiency of algorithms. Parallel programming techniques were used in this work to obtain better response times.

Our implementation supports multiple alignments of protein or nucleic acid sequences. Dynamic programming is used to identify identical residues among sequences. It is known that dynamic programming algorithms require a large amount of memory due to its quadratic space complexity. Therefore, optimization strategies will be studied and applied when aligning hundreds of sequences. MSA accuracy was evaluated through the sum of pairs scoring technique. The iterative process improves the MSA by applying a number of proposed operators. Different processes work independently searching for a better solution.

The comparison of the PaMSA algorithm with several common methods yielded promising results. Compared with the other methods tested (Muscle, Dialign and ClustalW), the PaMSA algorithm was less accurate, but the difference was not substantial. More work remains to be done to further improve accuracy. Despite the difference in accuracy, execution times achieved by PaMSA were better than the results obtained with the other methods tested.

Future work will focus on the study and application of optimization techniques in order to obtain more accurate alignments among sequences, and we will continue to apply parallel programming techniques to improve response time in algorithms used in MSA.

References

1. Altschul, S., Gish, W., Miller, W., Myers, E., Lipman, D.: Basic local alignment search tool. Molecular Biology-Elsevier 215(3), 403–410 (1990)
2. Anbarasu, L., Narayanasamy, P., Sundararajan, V.: Multiple molecular sequence alignment by island parallel genetic algorithm. Current Science 78(7), 858–863 (2000)
3. Bilu, Y., Agarwal, P., Kilodny, R.: Faster algorithms for optimal multiple sequence alignment based on pairwise comparisons. IEEE/ACM Transactions on Computational Biology and Bioinformatics 3(4), 408–422 (2006)
4. Chengpeng, B.: DNA motif alignment by evolving a population of Markov chains. BMC Bioinformatics 10(1), S13 (2009)
5. Edgar, R.: Muscle: multiple sequence alignment with high accuracy and high throughput. Nucleic Acids Research 32(5), 1792–1797 (2004)
6. Galperin, M., Cochrane, G.: The 2011 nucleic acids research database issue and the online molecular biology database collection. Nucleic Acids Research 39, D1–D6 (2011)
7. Gotoh, O.: Significant improvement in accuracy of multiple protein sequence alignments by iterative refinement as a assessed by reference to structural alignments. J. Mol. Biol. 264, 823–838 (1996)
8. Henikoff, S., Henikoff, J.G.: Amino acid substitution matrices from protein blocks. Biochemistry 89, 10915–10919 (1992)
9. Jones, N., Pevzner, P.A.: An introduction to bioinformatics algorithms. MIT Press (1996)
10. Kim, J., Pramanik, S., Chung, M.: Multiple sequence alignment using simulated annealing. Comput. Appl. Biosci. 10(4), 419–426 (1994)
11. Kleinjung, J., Douglas, N., Heringa, J.: Parallelized multiple alignment. Bioinformatics Applications Note 18(9), 1270–1271 (2002)
12. Lassmann, T., Frings, O., Sonnhammer, E.: Kalign2: high-performance multiple alignment of protein and nucleotide sequences allowing external features. Nucleid Acids Research 37(3), 858–865 (2009)
13. Li, K.: Clustalw-mpi: Clustalw analysis using distributed and parallel computing. Bioinformatics Applications Note 19(12), 1585–1586 (2003)
14. Lipman, D., Pearson, W.: Rapid and sensitive protein similarity searches. Science 227(4693), 1435–1441 (1985)
15. Lu, Y., Sze, S.: Improvig accuracy of multiple sequence alignment algorithms based on alignment of neighboring residues. Nucleic Acids Research 37(2), 463–472 (2009)
16. Luscombe, N., Greenbaum, D., Gerstein, M.: What is bioinformatics? a proposed definition and overview of the field. Method Inf. Med. 40(4), 346–358 (2001)
17. Moretti, S., Armougom, F., Wallace, I., Higgins, D., Jongeneel, C., Notredame, C.: The M-Coffee web server: a meta-method for computing multiple sequence alignments by combining alternative alignment methods. Nucleic Acids Research 35, Web Server Issue, W645–W648 (2007)
18. Mount, D.: Bioinformatics: sequence and genome analysis. Cold Spring Harbor Laboratory Press (2004)
19. National Center for Biotechnology Information: Fasta format, http://blast.ncbi.nlm.nih.gov/blastcgihelp.shtml
20. Needleman, S., Wunsch, C.: A general method applicable to the search for similarities in the amino acid sequence of two proteins. J. Mol. Biol. 48, 443–453 (1970)

21. Notredame, C., Higgins, D.: Saga: sequence alignment by genetic algorithm. Nucleic Acids Research 24(8), 1515–1524 (1996)
22. Notredame, C., Higgins, D., Heringa, J.: T-coffee: a novel method for fast and accurate multiple sequence alignment. J. Mol. Biol. 302(1), 205–217 (2000)
23. Shu, N., Elofsson, A.: KalignP: Improved multiple sequence alignments using position specific gap penalties in kalign2. Bioinformatics Applications Note 27(12), 1702–1703 (2011)
24. Smith, T., Waterman, M.: Identification of common molecular subsequences. J. Mol. Biol. 147, 195–197 (1981)
25. Thompson, J., Higgins, D., Gibson, T.: Clustal w: improving the sensitivy of progressive multiple sequence alignment through sequence weighting, position-specific gap penalties and weight matrix choice. Nucleic Acids Research 22(22), 4673–4680 (1994)
26. Wagner, R., Fischer, M.: The string-to-string correction problem. ACM 21(1), 168–173 (1974)
27. Wallace, I., O'Sullivan, O., Higgins, D., Notredame, C.: M-coffee: combining multiple sequence alignment methods with t-coffee. Nucleic Acids Research 34(6), 1692–1699 (2006)
28. Wang, Y., Li, K.: An adaptative and iterative algorithm for refining multiple sequence alignment. Computational Biology and Chemistry 28, 141–148 (2004)
29. Zhang, Z., Schwartz, S., Wagner, L., Miller, W.: A greedy algorithm for aligning dna sequences. Journal of Computational Biology 7(1/2), 203–214 (2000)

Author Index